Immunoassays in Food Analysis

Based on the Proceedings of a Symposium 'Immunoassays in Food Analysis', organised by M. N. Clifford and B. A. Morris for the Department of Biochemistry and the University of Surrey Food Science Group and held at the University of Surrey, Guildford, UK, 15–16 September 1983.

Chairmen:

Session I:
Professor V. Marks — *Division of Clinical Biochemistry, Department of Biochemistry, University of Surrey, Guildford, UK*

Session II:
Dr. H. W.-S. Chan — *AFRC Food Research Institute, Norwich, UK*

Session III:
Professor C. H. S. Hitchcock — *Unilever Research, Bedford, UK*

Immunoassays in Food Analysis

Edited by

B. A. MORRIS

and

M. N. CLIFFORD

Department of Biochemistry, University of Surrey, Guildford, UK

ELSEVIER APPLIED SCIENCE PUBLISHERS
LONDON and NEW YORK

ELSEVIER APPLIED SCIENCE PUBLISHERS LTD
Crown House, Linton Road, Barking, Essex IG11 8JU, England

Sole Distributor in the USA and Canada
ELSEVIER SCIENCE PUBLISHING CO., INC.
52 Vanderbilt Avenue, New York, NY 10017, USA

British Library Cataloguing in Publication Data

Immunoassays in food analysis.
1. Food—Analysis 2. Immunoassays
I. Morris, B. A. II. Clifford, M. N.
664'.07 TX545

ISBN 0-85334-321-7

WITH 29 TABLES AND 60 ILLUSTRATIONS

Photoset in Malta by Interprint Limited
Printed in Great Britain by Galliard (Printers) Ltd, Great Yarmouth

To Our Wives

Foreword

Throughout the last one hundred years a systematic approach to food analysis has been developed and documented by a number of professional bodies. This period has also been characterised by dramatic advances in food processing technology and by the increasing complexity of the food ingredients and processed foods presented for analysis. Accordingly, the food analyst has had to adopt and adapt techniques from diverse disciplines.

One such technique is that of immunoassay which utilises the exquisite specificity and sensitivity of the antigen–antibody reaction much used by clinical biochemists. This volume, which is based upon the papers presented at the first symposium on 'Immunoassays in Food Analysis', sets out the principles underlying the various antigen–antibody interactions, their specificity and precision, and describes a number of important applications for immunoassay in the field of food analysis.

Accordingly this volume is likely to be of considerable interest to analysts from academia, industry and the enforcement scene, and provides a sound basis for future developments.

R. Sawyer
Superintendent, Food and Nutrition Division,
Laboratory of the Government Chemist, London.

Preface

Immunoassay development and food analysis are two long-standing research interests of the Department of Biochemistry at the University of Surrey. Early in 1982, we realised that immunoassays could also become a very powerful tool for the food analyst, and that food analytes posed a number of problems previously unencountered by the immunoassayist. Accordingly, we began to apply the technique to this field of analysis.

We soon realised that while many organisations viewed this area of application with considerable interest, some were finding it difficult to progress because there was no reference book or training course specifically concerned with immunoassays in food analysis. This observation led us to organise the first ever symposium on this topic at the University of Surrey in September 1983. Approximately 80 people from the UK and Europe attended, the audience being drawn primarily from industry and analytical laboratories in more or less equal proportions.

This volume is based upon the papers presented at the symposium, with a valuable additional paper from Dr Johnston and his colleagues in Australia, who were unable to attend.

We should like to thank all the speakers, authors and chairmen for making it such a successful meeting; our colleagues for their critical contributions; Mrs Mary Lewis, our Symposium Secretary, for much effort on our behalf; and all the other University staff who helped in the organisation of the symposium and the production of this volume. We should particularly like to thank Mr G. B. Olley, of Elsevier Applied Science Publishers, for all his help and advice in the preparation of these proceedings for publication. We hope that it will enable a wider audience to become acquainted with immunoassay and benefit from the experiences of the pioneer workers in this particular field of application.

B. A. Morris
M. N. Clifford

Contents

Foreword vi

Preface vii

List of Contributors xi

Glossary xv
B. A. MORRIS

Session I: Principles of Immunoassay

1. The History of Immunoassays in Food Analysis . . . 3
M. N. CLIFFORD
2. Principles of Immunoassay 21
B. A. MORRIS
3. Principles of Enzyme Immunoassay 53
M. J. SAUER, J. A. FOULKES and B. A. MORRIS
4. Alternative Labels in Non-isotopic Immunoassay . . . 73
G. W. AHERNE

Session II: Application to Macromolecules

5. Species Identification of Meat in Raw, Unheated Meat Products 87
R. L. S. PATTERSON and S. J. JONES

6. Identification of the Species of Origin of Meat in Australia by Radioimmunoassay and Enzyme Immunoassay . . 95
L. A. Y. Johnston, P. D. Tracey-Patte, R. D. Pearson, J. G. R. Hurrell and D. P. Aitken
7. The Determination of Soya Protein in Meat Products (*Short Communication*) 111
C. H. S. Hitchcock
8. The Results of a Collaborative Trial to Determine Soya Protein in Meat Products by an ELISA Procedure (*Short Communication*) 113
R. Wood
9. Determination of Milk Protein Denaturation by an Enzyme-linked Immunosorbent Assay 115
L. M. J. Heppell
10. An Enzyme-linked Immunosorbent Assay for Amyloglucosidase in Beer 125
P. Vaag
11. Application of Enzyme Immunoassay Techniques for the Estimation of Staphylococcal Enterotoxins in Foods . . 141
P. D. Patel

Session III: Application to Small Molecules

12. An ELISA for the Analysis of the Mycotoxin Ochratoxin A in Food. 159
M. R. A. Morgan, R. McNerney and H. W.-S. Chan
13. The Use of Immunoassay for Monitoring Anabolic Hormones in Meat 169
M. J. Warwick, M. L. Bates and G. Shearer
14. Comparison of the Analysis of Total Potato Glycoalkaloids by Immunoassays and Conventional Procedures . . . 187
M. R. A. Morgan, R. McNerney, D. T. Coxon and H. W.-S. Chan
15. Cross-reactions in Immunoassays for Small Molecules: Use of Specific and Non-specific Antisera. 197
R. J. Robins, M. R. A. Morgan, M. J. C. Rhodes and J. M. Furze

Index 213

List of Contributors

G. W. Aherne

Division of Clinical Biochemistry, Department of Biochemistry, University of Surrey, Guildford, Surrey GU2 5XH, UK

D. P. Aitken

Immunochemistry Research and Development Section, Commonwealth Serum Laboratories, Parkville, Victoria, Australia 3052

M. L. Bates

Ministry of Agriculture, Fisheries and Food, Food Science Laboratory, Colney Lane, Norwich NR4 7UA, UK

H. W.-S. Chan

AFRC Food Research Institute, Colney Lane, Norwich NR4 7UA, UK

M. N. Clifford

Division of Nutrition and Food Science, Department of Biochemistry, University of Surrey, Guildford, Surrey GU2 5XH, UK

D. T. Coxon

AFRC Food Research Institute, Colney Lane, Norwich NR4 7UA, UK

J. A. Foulkes

Ministry of Agriculture, Fisheries and Food, Cattle Breeding Centre, Church Lane, Shinfield, Reading, Berks RG2 9BZ, UK

J. M. Furze

AFRC Food Research Institute, Colney Lane, Norwich NR4 7UA, UK

L. M. J. Heppell

Nutrition Department, National Institute for Research in Dairying, Shinfield, Reading, Berks RG2 9AT, UK. Present address (from 1 April 1985): *Animal and Grassland Research Institute, Shinfield, Reading, Berks RG2 9AT, UK.*

C. H. S. Hitchcock

Unilever Research, Colworth Laboratory, Colworth House, Sharnbrook, Bedford MK44 1LO, UK

J. G. R. Hurrell

Immunochemistry Research and Development Section, Commonwealth Serum Laboratories, Parkville, Victoria, Australia 3052

L. A. Y. Johnston

CSIRO, Division of Tropical Animal Science, Long Pocket Laboratories, PMB No. 3, Indooroopilly, Queensland, Australia 4068

S. J. Jones

AFRC Meat Research Institute, Langford, Bristol BS18 7DY, UK

R. McNerney

AFRC Food Research Institute, Colney Lane, Norwich NR4 7UA, UK

M. R. A. Morgan

AFRC Food Research Institute, Colney Lane, Norwich NR4 7UA, UK

B. A. Morris

Division of Clinical Biochemistry, Department of Biochemistry, University of Surrey, Guildford, Surrey GU2 5XH, UK

P. D. Patel

Microbiology Section, Leatherhead Food Research Association, Randalls Road, Leatherhead, Surrey KT22 7RY, UK

R. L. S. Patterson

AFRC Meat Research Institute, Langford, Bristol BS18 7DY, UK

R. D. Pearson

CSIRO, Division of Tropical Animal Science, Long Pocket Laboratories, PMB No. 3, Indooroopilly, Queensland, Australia 4068

M. J. C. Rhodes

AFRC Food Research Institute, Colney Lane, Norwich, NR4 7UA, UK

R. J. Robins

AFRC Food Research Institute, Colney Lane, Norwich NR4 7UA, UK

M. J. Sauer

Ministry of Agriculture, Fisheries and Food, Cattle Breeding Centre, Church Lane, Shinfield, Reading, Berks RG2 9BZ, UK

G. Shearer

Ministry of Agriculture, Fisheries and Food, Food Science Laboratory, Colney Lane, Norwich NR4 7UA, UK

P. D. TRACEY-PATTE

CSIRO, Division of Tropical Animal Science, Long Pocket Laboratories, PMB No. 3, Indooroopilly, Queensland, Australia 4068

P. VAAG

Department of Biotechnology, Carlsberg Research Laboratory, DK-2500 Valby, Copenhagen, Denmark

M. J. WARWICK

Ministry of Agriculture, Fisheries and Food, Food Science Laboratory, Colney Lane, Norwich NR4 7UA, UK

R. WOOD

Ministry of Agriculture, Fisheries and Food, 65 Romney Street, London SW1 3RD, UK

Glossary

(compiled by B. A. Morris)

Analyte. The substance in a test sample whose quantity is to be determined or presence detected.

Antibody or immunoglobulin. A binding protein which is synthesised by the immune system of an animal in response to either the invasion of a foreign organism or the injection of an immunogen. May be divided into a number of different classes, the principal one which is usually employed in an immunoassay being immunoglobulin G, IgG (MW 160 000) (cf. Antiserum).

Antigen. A substance that will react with its specific antibody (cf. Hapten, Immunogen).

Antigen-free matrix. Matrix (q.v.) from which the antigen has been selectively removed, e.g. by an immunosorbent, for the purpose of diluting the standards.

Antigenic determinant. Feature of an antigen which defines the recognition pattern of an antibody.

Antiserum. Serum containing antibodies (q.v.) to a specified antigen.

Avidity. The energy with which the combining sites of an antibody bind its specific antigen. It is essentially the same as the association constant (K_A) in physical chemistry with

$$K_A = \frac{[AgAb]}{[Ab][Ag]} \text{ in litres/mole,}$$

where [AgAb], [Ab] and [Ag] are the molar concentrations of the antigen–antibody complex, free antibody and free antigen respectively.

Note: The terms *avidity* and *affinity* are often used synonymously and both relate to the energy of binding of a particular antigen–

antibody combination. Strictly speaking, however, the term 'avidity' refers specifically to the properties of an antibody, and the term 'affinity' to those of the antigen. Numerically, however, the avidity of the antibody for the antigen is equal to the affinity of the antigen for the antibody.

Bias. A systematic error in the assay system.

Bound fraction. The portion of the incubation mixture which contains the antigen–antibody complex (cf. Free fraction).

Carrier protein. A large molecular weight protein to which haptens (q.v.) are covalently linked in order to elicit an immune response to the latter.

Classical immunoassay. The original immunoassay concept, first reported by Yalow & Berson (1960), is a limited reagent immunoassay (q.v.) in which antigen (the analyte) and labelled antigen compete for a fixed, but limited, number of antibody binding sites (cf. Immunometric assay).

Competitive ELISA. A limited reagent form of enzyme-linked immunosorbent assay, in which the antigen in a test sample or standard solution competes with enzyme-labelled antigen for the limited binding sites on the immobilised antibody fraction (cf. Non-competitive ELISA).

Cross-reaction. Ability of substances, other than the antigen, to bind to the antibody, and the ability of substances other than the antibody to bind the antigen. Such substances, if present in a test sample, may compete with the antigen for the binding site, thus leading to an erroneous potency estimate. These substances may be natural precursors of the antigen (or binding protein), degradation products (from degradation *in vivo* or *in vitro*) or other substances that carry on their surface a molecular configuration similar to the antigenic determinants on the antigen being measured.

Delayed addition immunoassay. A limited reagent immunoassay in which the antigen (in the test sample or standard) is incubated with antibody for a period prior to the addition of labelled antigen. If the antibody is of sufficiently high avidity, this form of assay may increase sensitivity, compared with the more usual assay procedure in which all three reactants are incubated together for the same length of time.

Detection limit. The smallest amount or concentration of analyte which, with a stated confidence (commonly two standard deviations, or expressed as confidence or fiducal limits), can be distinguished from zero. This value depends on the precision of the measurements of zero dose solution and of the specimen.

Disequilibrium (or non-equilibrium) immunoassay. Immunoassay in which the reaction between antigen and antibody is stopped before equilibrium has been reached. Frequently used in continuous flow automated immunoassay systems which enable exact reproducible timing between reagent addition and the separation step (cf. Equilibrium immunoassay).

Enzyme immunoassay. Sometimes referred to as *Enzymoimmunoassay*. An assay procedure based on the reversible and non-covalent binding of an antigen by a specific antibody, in which one of the reactants is labelled with an enzyme.

Enzyme-linked immunosorbent assay (ELISA). An enzyme immunoassay (q.v.) in which one of the reactants is adsorbed on to the surface of the wells of a microtitre plate (q.v.).

Enzyme mediated immunoassay technique (EMIT). A homogeneous enzyme immunoassay (q.v.) for haptens (q.v) in which the enzyme is linked to the hapten in such a way that the enzyme activity is altered when the hapten combines with its antibody. The activity may be enhanced or, more usually, reduced. EMIT is the Registered Trade Mark of the Syva Corp., USA.

Equilibrium immunoassay. An immunoassay in which the assay components in the incubation mixture are allowed to react until the concentration of each reactant has ceased to change with time, i.e. equilibrium has been attained (cf. Delayed addition immunoassay; Disequilibrium immunoassay).

Excess reagent immunoassay. An immunoassay in which the antibody is present in excess, as in immunometric assays (q.v.) (cf. Limited reagent immunoassay).

First antibody. An antibody reacting specifically with the antigen being measured (cf. Second antibody).

Fluoroimmunoassay. A classical immunoassay (q.v.) in which the antigen is labelled with a fluorophore for use as the tracer.

Free fraction. The portion of the incubation mixture which, after phase separation (q.v.), does not contain the antigen–antibody complex (cf. Bound fraction). It may be free antigen, as in classical immunoassay (q.v.), or free antibody, as in immunometric assay (q.v.).

Hapten. An antigen (q.v.) which is not usually immunogenic but which becomes so when coupled to a larger molecule, usually a protein (cf. Immunogen).

Heterogeneous immunoassay. An immunoassay in which it is necessary to separate the antigen–antibody complex (bound fraction, q.v.) from the free reactants, usually antigen, prior to measuring the quantity of label in either the bound or free fraction (cf. Homogeneous immunoassay).

Homogeneous immunoassay. An immunoassay in which it is not necessary to separate the antigen–antibody complex from the free reactants prior to end-point measurement (cf. Heterogeneous immunoassay).

Hook effect. A phenomenon associated with the standard curve of a classical immunoassay in which increasing the concentration of antigen from zero level initially produces an *increase* in binding of the labelled antigen before resulting in a fall in binding, i.e. a 'hooked' curve; the problem may be overcome by using the antiserum at a higher dilution.

Immunoassay. An assay procedure based on the reversible and non-covalent binding of an antigen by antibody using a labelled form of one or the other to quantify the system. Can be used to detect or quantify either antigens or antibodies.

Immunogen. A substance that, when injected into a suitable animal, stimulates the production of antibody or antibodies that can combine with the same substance as an antigen (q.v.).

Immunometric assay. A non-competitive excess reagent immunoassay (q.v.) based on the reversible and non-covalent binding of an antigen by excess specific antibody (or antibodies) labelled with a tracer molecule. An immunoassay using labelled antibodies instead of labelled antigens to quantitate the assay. (See Sandwich or two-site assays.)

Immunoreactivity. The ability of a specified antigen to combine with its antibody, or a specified antibody to combine with its antigen.

Inaccuracy. The numerical difference between the average of a series of estimates and the true or accepted value.

Label. The substance attached to one of the assay reactants for the purpose of determining the proportion of that reactant which has formed an antigen–antibody complex, by yielding a perceptible signal.

Labelled antigen. A form of the specified antigen to which a label (q.v.) has been attached covalently.

Limited reagent immunoassay. An immunoassay in which the amount of antibody added is insufficient to bind all the labelled antigen in the reaction mixture (cf. Excess reagent immunoassay).

Luminoimmunoassay. A classical immunoassay (q.v.) in which the antigen is labelled with either a bioluminescent or chemiluminescent molecule for use as the tracer.

Matrix. Substances other than the analyte which are present in the sample or sample extract (cf. Antigen-free matrix).

Microtitre plate. A disposable plastic tray, containing 96 wells, which is used to hold the incubates in ELISA (q.v.).

Misclassification. The extent to which the free, i.e. non-bound, reactants are present in the bound fraction and vice versa after phase separation (q.v.).

Monoclonal antibodies. Antibodies derived from a single clone of lymphocytes, produced by a hybridoma as a result of fusion of a sensitised lymphocyte with a myeloma cell (cf. Polyclonal antibodies).

Multivalent antigen. An antigen with two or more antigenic determinants (q.v.).

Non-competitive ELISA. An excess reagent form of enzyme-linked immunosorbent assay (cf. Competitive ELISA).

Non-specific binding (NSB). The fraction of labelled material present in the apparent bound fraction for reasons other than specific binding to the binding site of an antibody.

Parallelism. The extent to which the dose-response curves of two substances are identical, except for displacement along the dose axis of one relative to the other. If the curves are curvilinear, this condition is described as 'generalised parallelism'. It is one test of identity of two preparations, e.g. analyte and standard.

Phase separation. Procedure by which the free fraction is separated from the bound fraction prior to measurement of the amount of labelled reactant present in one or the other fraction.

Polyclonal antibodies. Antibodies to a given antigen which are derived from several clones of lymphocytes produced in a single animal as the result of an injection of immunogen (cf. Monoclonal antibodies).

Precision/imprecision. The closeness of agreement between the results obtained by applying a given procedure several times under prescribed conditions.

As precision has no numerical value, the term imprecision, to which a numerical value can be assigned, may be preferable in some contexts.

Imprecision consists of variability among replicate measurements of the same material, and is usually expressed as the standard deviation, or variance or coefficient of variation.

Precision profile. Graphical representation of the precision, or more correctly, the imprecision, of measurements obtained with an assay system over a range of dose levels of the analyte.

Such profiles are of value in assessing the reliance that can be placed on an assay estimate at a particular dose level, because the precision in immunoassays may vary considerably with analyte concentration.

Radioimmunoassay. A classical immunoassay (q.v.) in which the antigen is labelled with a radionucleide for use as the tracer.

Ruggedness. Characteristic of an assay system which makes the results obtained unaffected by changes in the assay reagents and procedures.

In practice, non-ruggedness is manifested by poor precision, poor inter-assay variability and poor interlaboratory agreement.

Sandwich or two-site assay. An immunometric assay (q.v.) using two different antibodies, each recognising a separate antigenic determinant (q.v.) on the analyte molecule, one of the antibodies being immobilised on a solid support and the other being labelled with a tracer molecule. By sequential incubations, the analyte is sandwiched between the two antibodies. This assay system is inherently more specific than classical immunoassay (q.v.), since two separate antigenic determinants on the analyte are recognised simultaneously. Its use is obviously restricted to the measurement of multivalent antigens (q.v.).

Second or double antibody. An antibody raised against the immunoglobulins (usually IgG) of the species in which the first antibody (q.v.) was raised.

Sensitivity. The ability of a method to distinguish significantly between small differences in concentration of analyte.

Specificity. The specificity of an antibody is the degree to which it is not influenced by substances of similar structure and composition.

Titre. The final dilution of an antiserum or purified antibody preparation which binds a specified fraction (commonly 50%), or a specified amount, of a stated quantity of the labelled antigen: usually expressed numerically as the reciprocal of the dilution.

Tracer. See Label.

Two-site assay. See Sandwich assay.

REFERENCES AND BIBLIOGRAPHY

SÖNKSEN, P. H. (1974) Radioimmunoassay and saturation analysis. *British Medical Bulletin*, **30**, 103.

WHO Expert Committee on Biological Standardisation (1981) Requirements for Immunoassay Kits, World Health Organisation Technical Report, Series No. 658, WHO, Geneva.

YALOW, R. S. & BERSON, S. A. (1960) Immunoassay of endogenous plasma insulin in man. *Journal of Clinical Investigation*, **39**, 1157–75.

SESSION I

Principles of Immunoassay

1

The History of Immunoassays in Food Analysis

M. N. CLIFFORD

Division of Nutrition and Food Science, Department of Biochemistry, University of Surrey, Guildford, UK

INTRODUCTION

The objective of this first chapter is to trace the development of the food immunoassay, a term which embraces any immunoassay applied to any material that is destined for incorporation in or for use as human food. In practice the majority of such assays are found most conveniently in *Food Science and Technology Abstracts* (*FSTA*) since *Immunoassay Abstracts* has overlooked publications concerned with food analysis. However this does not imply that immunoassay papers reported elsewhere are irrelevant to food immunoassayists.

EARLY DEVELOPMENTS

Ultimately, all developments in immunoassay stem from the report of the first immunoassay by Yalow & Berson in 1959. Their immunoassay was for insulin, employed a radiolabelled form of the hormone, and was able to detect some 10 to 40 pg of insulin. This radioimmunoassay (RIA) doubtless appeared as an esoteric research tool in 1959 but such assays were rapidly transformed into routine quantitative methods for many analytes (e.g. see Sonksen, 1974) and because such RIAs provided superior sensitivity, greater specificity and higher throughput than traditional methods of analysis, they gained rapid and widespread acceptance among clinical scientists.

In contrast, food scientists were slow to adopt and adapt such assays. The first food immunoassay seems to have been reported by Porath

(1970) at the International Biological Programme in Stockholm in 1968. This brief paper recommended the use of RIAs for the detection of specific proteins in food extracts at concentrations below 1 pg/ml. Figure 1 shows that further publications in this field were slow to appear, the total reaching only seven by 1973.

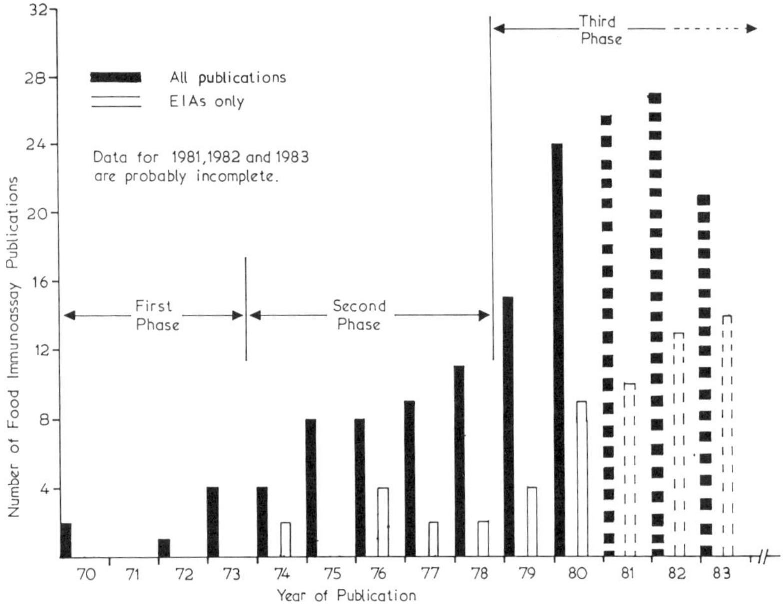

Fig. 1. The growth in food immunoassays illustrating the recent impact of EIAs.

Several reasons can be suggested for this slow transfer of technology:

(a) conservatism among food analysts;
(b) high overall initial cost because of the need for special equipment, special facilities for the handling and disposal of radioisotopes, and training of personnel; and in particular:
(c) difficulties in dealing with solid samples;
(d) recognition of lay concern, not to say hysteria, at the thought of radioactivity in close proximity to human food.

The means to circumvent the last two objections, and thus potentially the first objection also, was provided in 1971 with simultaneous reports of a major development (Engvall & Perlmann, 1971; van Weemen &

Schuurs, 1971). These workers showed that radiolabels could be replaced by enzyme labels and thus the radioisotopes and their attendant problems could be dispensed with once the enzyme immunoassay (EIA) had been developed. Reviews by Voller *et al.* (1979) and Voller & Bidwell (1980) illustrate the routine adoption of EIAs in clinical analysis.

DEVELOPMENT OF ENZYME IMMUNOASSAYS

Probably the first reports of food EIAs can be attributed to Engvall and colleagues in 1974 (Ljungström *et al.*, 1974; Ruitenberg *et al.*, 1974). These publications, which were concerned with the detection of parasites in pigs for slaughter, triggered the start of the second phase in the development of the food immunoassays. Figure 1 shows that during this phase there were, on average, two publications per year describing food EIAs and that these represented some 25% of the total number of publications describing food immunoassays. During this phase there began a

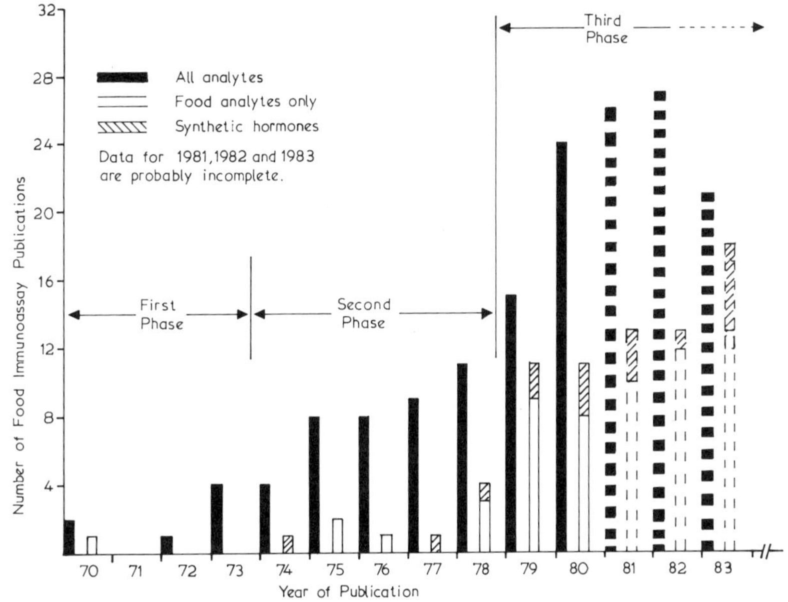

Fig. 2. The growth in food immunoassays illustrating the progressive move from traditional to food analytes.

broadening in the type of analysis for which assays were developed. This is illustrated in Fig. 2 which compares the number of publications concerned with traditional analytes and food analytes. The division between these types is somewhat arbitrary and not always sharp. Traditional analytes have been defined as analytes similar to those routinely assayed in clinical applications and a brief list is given in Table 1. However, the substantial category 'hormones' includes 17 publications concerned with synthetic analogues rather than natural compounds. These assays clearly have a clinical origin but could be considered as assays for food contaminants and as such appear as a distinct feature of the histograms for food analytes shown in Fig. 2. During phase two, publications concerned with food analytes accounted for only 15% of the total, or 22% if the synthetic hormones are included.

Table 1
Food Immunoassay Publications Classified by Type of Analytes

Type of analyte	*Number of publications*	
Traditional analytes		
Bacteria and bacterial toxins	42	
Hormones (including 17 papers on synthetic analogues)	30	
Mycotoxins	20	
Parasites	7	
Viruses	3	
Others	1	
Total		103
Food analytes		
Proteins (excluding toxins)	20	
Other food components, e.g. caffeine, ciguatoxin, glycoalkaloids, limonin, naringin and vitamins	22	
Additives and contaminants, e.g. antibiotics, fungicides, surfactants, but not anabolic steroids	13	
Total		55
General papers		2
Total publications		160

RECENT DEVELOPMENTS

There was no distinct trigger to initiate the third phase, rather an intensification of the trend that began in the second phase consequent upon the general recognition that food was not just another substrate to which some traditional assays could be applied but also a complex matrix presenting its own challenges and opportunities to the immunoassayist. This maturation is clearly illustrated in Figs 1 and 2. Even though the more recent data must be treated as provisional, the essentially exponential trend appears to be continuing. Averaged over the whole of phase three, up to December 1983, publications reporting EIAs represent 45% of the total; food analytes represent 45 or 56% depending on whether or not the synthetic hormones are included.

CONCLUSIONS

In addition to the traditional and food analytes listed in Table 1, the indexed bibliography which follows includes publications concerned with the detection and/or quantification of insect parts, a class of odour impact compounds, a range of sweet-tasting compounds (including ranking in agreement with their sweetness judged organoleptically) and the judgement of citrus fruit authenticity. This considerable spread of analytes gives an exciting but realistic glimpse of the tasks, often difficult with conventional techniques, to which food immunoassays may be applied with an expectation of success.

The well established exponential growth of interest in food immunoassays is most welcome, but the cynic could liken this to a state of euphoria and would say that euphoria is but one step away from disillusion. This has not happened in clinical analysis and need not in food analysis. The major objectives of the chapters which follow are to show what has recently been achieved and to present and explain concisely the principles which must be followed to produce successful assays and thus ensure a healthy future for immunoassays in food analysis.

ACKNOWLEDGEMENT

I wish to record the considerable assistance that my colleague Brian Morris has given in facilitating my entry into the field of immunoassays.

REFERENCES

ENGVALL, E. & PERLMANN, P. (1971) Enzyme-linked immunosorbent assay (ELISA). Quantitative assay of immunoglobulin G. *Immunochemistry*, **8**, 871–4.

LJUNGSTRÖM, I., ENGVALL, E. & RUITENBERG, E. J. (1974) ELISA — Enzyme-linked immunosorbent assay — in serological diagnosis of *Trichinella spiralis* infection. *Proceedings of Third International Congress of Parasitology*, Vols I, II, III, World Federation of Parasitologists, Congress Centre, Exhibition Grounds, Munich, Aug. 25–31, Facta Publications, Vienna, pp. 1204–5.

PORATH, J. (1970) Radioimmunoassays as a means of detecting small quantities of specific proteins. In: *Evaluation of Novel Protein Products*, Bender, A. E., Kihlberg, R., Loefqvuist, B. & Munck, L. (eds), Pergamon Press, Oxford, pp. 285–6.

RUITENBERG, E. J., STEERENBERG, P. A., BROSI, B. J. M., BUYS, J., LJUNGSTRÖM, I. & ENGVALL, E. (1974) Application of ELISA for the serodiagnosis of *T. spiralis* infections in pigs under slaughterhouse conditions. *Third International Congress of Parasitology*, World Federation of Parasitologists, Facta Publications, Vienna, pp. 1203–4.

SONKSEN, P. H. (1974) Radioimmunoassay and saturation analysis. *British Medical Bulletin*, **30**, 1–103.

VAN WEEMÉN, B. K. & SCHUURS, A. H. W. M. (1971) Immunoassay using antigen–enzyme conjugates. *FEBS Letters*, **15**, 232–6.

VOLLER, A., BIDWELL, D. E. & BARTLETT, A. (1979) *The Enzyme Linked Immunosorbent Assay (ELISA)*, Vol. 1, Dynatech Europe, Guernsey.

VOLLER, A. & BIDWELL, D. (1980) *The Enzyme Linked Immunosorbent Assay (ELISA)* Vol. 2, Dynatech Europe, Guernsey.

YALOW, R. S. & BERSON, S. A. (1959) Assay of plasma insulin in human subjects by immunological methods. *Nature*, **184**, 1643–4.

BIBLIOGRAPHY

These papers have been collected primarily from *Food Science and Technology Abstracts* up to and including December 1983. Some more recent papers have been collected from the original literature. The data for 1981, 1982 and 1983 will be incomplete.

1. AALUND, O., BRUNFELDT, K., HALD, B., KROGH, P. & POULSEN, K. (1975) A radioimmunoassay for Ochratoxin A: a preliminary investigation. *Acta Pathologica Microbiologica Scandinavica Section C*, **83**, 390–2.
2. ACKERMAN, J. I. & CHESBRO, W. (1981)* Detection of *Staphylococcus*

*Note: This paper has also been cited as 1980 probably because this date appears on the journal cover whereas 1981 appears on the title pages of all papers in this volume.

aureus products in foods using enzyme-linked immunosorbent assay and spectrophotometric thermonuclease assay. *Journal of Food Safety*, **3** (1), 15–25.

3. AITKEN, D. P., CHANDLER, H. M., PREMIER, R. & HURRELL, J. G. R. (1983) Meat species identification using urease-antibody. In: *Horizon 90: How to survive the 80s. 16th Annual Convention Australian Institute of Food Science and Technology*, p. 17.
4. AL-RUBAE, A. Y. (1979) The enzyme-linked immunosorbent assay: a new method for the analysis of pesticide residues. *Dissertation Abstracts International B*, **39** (10), 4723–4.
5. ANDERSON, M. (1979) Enzyme immunoassay for measuring lipoprotein lipase activator in milk. *Journal of Dairy Science*, **62** (9), 1380–3.
6. ARESON, P. D. W., CHARM, S. E. & WONG, B. L. (1980) Determination of staphylococcal enterotoxins A and B in various food extracts, using staphylococcal cells containing protein A. *Journal of Food Science*, **45** (2), 400–1.
7. ARNSTADT, K. I. (1981) Enzyme immunoassay (EIA) for the synthetic oestrogen diethylstilboestrol (DES). *Zeitschrift für Lebensmittel-Untersuchung und -Forschung*, **173** (4), 255–60.
8. BENNETT, R. W., KEOSEYAN, S. A., TATINI, S. R., THOTA, H., & COLLINS, W. S. (1973) Staphylococcal enterotoxin: a comparative study of serological detection methods. *Canadian Institute of Food Science and Technology Journal*, **6** (2), 131–4.
9. BERGDOLL, M. S. & REISER, R. (1980) Application of radioimmunoassay for detection of staphylococcal enterotoxins in foods. *Journal of Food Protection*, **43** (1), 68–72.
10. BERGER, L. R. & BERGER, J. A. (1978) Development of a colorimetric assay for ciguatoxin in fish muscle. *Abstract of the Annual Meeting of The American Society for Microbiology*, **78**, 194.
11. BIANCIFIORI, F., GIALLETTI, L., FRESCURA, T. & MOROZZI, A. (1981) Use of protein A for indirect serological diagnosis of porcine trichinosis. *Archivio Veterinario Italiano*, **32** (5/6), 143–7.
12. BIERMANN, A. & TERPLAN, G. (1980) Aflatoxin B_1 detection by ELISA. *Archiv für Lebensmittelhygiene*, **31** (2), 51–7.
13. BIERMANN, A & TERPLAN, G. (1982) Determination of aflatoxin B_1 in foods by micro-ELISA. *Archiv für Lebensmittelhygiene*, **33** (1), 17–20.
14. BOKX, J. A. DE, PIRON, P. G. M. & MAAT, D. Z. (1980) Detection of potato virus X in tubers with enzyme-linked immunosorbent assay (ELISA). *Potato Research*, **23** (1), 129–31.
15. BONNEAU, M. & DESMOULIN, B. (1980) Backfat androstenone content in entire male pigs of the Large White breed: variations according to social conditions during rearing. *Reproduction, Nutrition, Développement*, **20** (5A), 1429–32.
16. BONNEAU, M., DESMOULIN, B., FROUIN, A. & BIDARD, J.-P. (1980) Effects of processing of boar meat on the androstenone content of the fat. *Annales de Technologie Agricole*, **29** (1), 69–73.
17. BRUNN, H. (1981) Detection of diethyl stilboestrol in trout. *Archiv für Lebensmittelhygiene*, **32** (5), 144.

18. BRUNN, H., STOJANOWIC, V., FISTER, P., EISENBRODT, E. (1983) False positive radioimmunoassay results obtained during the detection of stilbene derivatives after using fluorinated corticosteroids. *Fleischwirtschaft*, **63** (3), 398–400.
19. BRUNN, H., STOJANOWIC, V., FISTER, P., EISENBRODT, E. (1983) Metabolites of fluorinated corticosteroids as the cause of false positive radioimmunoassay results during the detection of stilbene derivatives. *Fleischwirtschaft*, **63** (3), 401–5.
20. BUENING-PFAUE, H., TIMMERMANNS, P. & NOTERMANS, S. (1981) A simple method for detection of staphylococcal enterotoxin B in vanilla custard by the ELISA test. *Zeitschrift für Lebensmittel-Untersuchung und Forschung*, **173** (5), 351–5.
21. BUKOVIC, J. A. & JOHNSON, H. M. (1975) Staphylococcal enterotoxin C: solid-phase radioimmunoassay. *Applied Microbiology*, **30** (4), 700–1.
22. BUTLER, W. R. & BORDES, C. K. DES (1980) Radioimmunoassay technique for measuring cortisol in milk. *Journal of Dairy Science*, **63** (3), 474–7.
23. CLAUS, R. (1975) Determination of boar taint in the fat of swine by a radioimmunological method. I. Accumulation of boar taint in relation to the age of the boar. *Zeitschrift für Tierzüchtung und Züchtungsbiologie*, **92** (1/2), 118–26.
24. CLAUS, R. (1976) Determination of the boar taint compound in pork fat by a radioimmunological method. II. The time course of clearance of boar taint after castration. *Zeitschrift für Tierzüchtung und Züchtungsbiologie*, **93** (1), 38–47.
25. CLAUS, R. & HOPPEN, H. O. (1979) The boar-pheromone steroid identified in vegetables. *Experientia*, **35** (12), 1674–5.
26. CLAUS, R., HOPPEN, H. O. & KARG, H. (1981) The secret of truffles: a steroidal pheromone? *Experientia*, **37** (11), 1178–9.
27. COLLINS, W. S., JOHNSON, A. D., METZGER, J. F. & BENNETT, R. W. (1973) Rapid solid-phase radioimmunoassay for staphylococcal enterotoxin A. *Applied Microbiology*, **25** (5), 774–7.
28. COXON, D. T. (1984) Methodology for glycoalkaloid analysis. *American Potato Journal*, **61**, 169–83.
29. CRIMES, A. A.,* BAILEY, F. J. & HITCHCOCK, C. H. S. (1981) Determination of foreign proteins in meat products. *Analytical Proceedings*, **18** (4), 164–6.
30. CROSBY, N. T. (1983) Enzymes in food analysis. *Trends in Analytical Chemistry*, **2** (1), VII–VIII.
31. DARRAGH, A., LAMBE, R. F., HALLINAN, D. & O'KELLY, D. A. (1979). Caffeine in soft drinks. *Lancet*, **1**, 1196.
32. DICKIE, N. (1970). Detection of staphylococcal enterotoxin. *Canadian Institute of Food Technology Journal*, **3** (4), 143–4.
33. DICKIE, N. & AKHTAR, S. M. (1982) Improved radioimmunoassay of staphylococcal enterotoxin A. *Journal of the Association of Official Analytical Chemists*, **65** (1), 180–4.

*This citation is correct. Note that this paper has also been abstracted incorrectly as Grimes, A. A.

34. DONHAUSER, S. (1979) Radioimmunoassay of bromelin and ficin in beer. *Brauwissenschaft*, **32** (7), 211–13.
35. EL-NAKIB, O., PESTKA, J. J. & CHU, F. S. (1981) Determination of aflatoxin B_1 in corn, wheat, and peanut butter by enzyme-linked immunosorbent assay and solid phase radioimmunoassay. *Journal of the Association of Official Analytical Chemists*, **64** (5), 1077–82.
36. ERCEGOVICH, C. D., VALLEJO, R. P., GETTIG, R. R., WOODS, L., BOGUS, E. R. & MUMMA, R. O. (1981) Development of a radioimmunoassay for parathion. *Journal of Agricultural and Food Chemistry*, **29** (3), 559–63.
37. EVERETT, A. W., PRIOR, G., CLARK, W. A., ZAK, R. (1983) Quantitation of myosin in muscle. *Analytical Biochemistry*, **130** (1), 102–7.
38. FAULSTICH, H. & COCHET-MEILHAC, M. (1976) Amatoxins in edible mushrooms. *FEBS Letters*, **64** (1), 73–5.
39. FEY, H., STIFFLER-ROSENBERG, G., WARTENWEILER-BURKHARD, G., MÜELLER, C., RÜEESS, O. (1982) Detection of staphylococcal enterotoxins. *Schweizer archiv für Tierheilkunde*, **124** (6), 297–306.
40. FIRON, N., LIFSHITZ, A., RIMON, A. & HOCHBERG, Y. (1979) An immunoassay method for estimating the orange juice content of commercial soft drinks. *Lebensmittel-Wissenschaft und-Technologie*, **12** (3), 143–6.
41. FOARD, D. E., HWANG, D. L., PAO, W. L., WONG-HUANG, W. O. & YANG, W. K. (1979) Radioimmunoassay of soybean proteins having high methionine, high half-cystine content. Quantitative variation in seeds of M3 plants and of a sampling of the World collection. In: *Seed Protein Improvement in Cereals and Grain Legumes*, Vol. 11, International Atomic Energy Agency, FAO, p. 430.
42. FOGLESONG, M. A. & LE FEBER, D. S. (1982) Radioimmunoassay for Hygromycin B in feeds. *Journal of the Association of Official Analytical Chemists*, **65** (1), 48–51.
43. FORLAND, D. M., LUNDSTROM, K. & ANDRESEN, O. (1980) Relationship between androstenone content in fat, intensity of boar taint and size of accessory sex glands in boars. *Nordisk Veterinaermedicin*, **32** (5), 201–6.
44. FREED, R. C., EVENSON, M. L., REISER, R. F. & BERGDOLL, M. S. (1982) Enzyme-linked immunosorbent assay for detection of staphylococcal enterotoxins in foods. *Applied and Environmental Microbiology*, **44** (6), 1349–55.
45. GOREWIT, R. C. (1978) Radioimmunoassay of oxytocin in milk. *Journal of Dairy Science*, **61** (Suppl. 1), 148–9.
46. GRIDLEY, J. C., ALLEN, E. H. & SHIMODA, W. (1983) Radioimmunoassay for diethylstilboestrol and the monoglucuronide metabolite in bovine liver. *Journal of Agricultural and Food Chemistry*, **31** (2), 292–6.
47. GROHMANN, H. & STAN, H.-J. (1982) Specific determination of oestrogen-active anabolic agents in meat by the oestrogen receptor test and radioimmunoassay after HPLC separation. *Lebensmittelchemie und Gerichtliche Chemie*, **36** (2), 34–6.
48. GRUNERT, E., SCHULTE, B. & SCHULTZ, G. (1974) Possibilities and limits of detection of oestrogens after treatment of cattle. *Ubersichten zur Tierernährung*, **2** (2), 135–55.

49. GUENTHER, H. O. (1982) Detection of natural oestrogens by enzyme immunoassay. *Lebensmittelchemie und Gerichtliche Chemie*, **36** (4), 97.
50. GUGERLI, P. & GEHRIGER, W. (1980) Enzyme-linked immunosorbent assay (ELISA) for the detection of leafroll virus and potato virus Y in potato tubers after artificial break of dormancy. *Potato Research*, **23** (3), 353–9.
51. HARDER, W. O. & CHU, F. S. (1979) Radioimmunoassay of aflatoxin M_1. *Abstracts of the Annual Meeting of the American Society for Microbiology*, **79**, 204.
52. HARDER, W. O. & CHU, F. S. (1979) Production and characterization of antibody against aflatoxin M_1. *Experientia*, **35** (8), 1104–7.
53. HEBEL, R.-D. (1975) Radioimmunoassay: method and application in food inspection. *Fleischwirtschaft*, **55** (6), 812–13.
54. HENRICKS, D. M. & TORRENCE, A. K. (1978) Endogenous estradiol-17β in bovine tissues. *Journal of the Association of Official Analytical Chemists*, **61** (5), 1280–3.
55. HITCHCOCK, C. H. S., BAILEY, F. J., CRIMES, A. A., DEAN, D. A. G. & DAVIS, P. J. (1981) Determination of soya proteins in food using an enzyme-linked immunosorbent assay (ELISA) procedure. *Journal of the Science of Food and Agriculture*, **32** (2), 157–65.
56. HOFFMANN, B. (1978) Use of radioimmunoassay for monitoring hormonal residues in edible animal products. *Journal of the Association of Official Analytical Chemists*, **61** (5), 1263–73.
57. HOFFMANN, B. & LASCHUETZA, W. (1980). Radioimmunoassay for determining diethylstilboestrol in blood plasma and edible parts of cattle. *Archiv für Lebensmittelhygiene*, **31** (3), 105–11.
58. HOFFMANN, B. & BLIETZ, C. (1983) Application of radioimmunoassay (RIA) for the determination of residues of anabolic sex hormones. *Journal of Animal Science*, **57** (1), 239–46.
59. HOLLIS, B. W., ROOS, B. A., DRAPER, H. H. & LAMBERT, P. W. (1981) Vitamin D and its metabolites in human and bovine milk. *Journal of Nutrition*, **111** (7), 1240–8.
60. HOUGH, C. A. M. & EDWARDSON, J. A. (1978) Antibodies to thaumatin as a model of the sweet taste receptor. *Nature*, **271**, 381–3.
61. IKI, K., SEKIGUCHI, K., KURATA, K., TADA, T., NAKAGAWA, H., OGURA, N. & TAKEHANA, H. (1978) Immunological properties of β-fructofuranosidase from ripening tomato fruit. *Phytochemistry*, **17** (2), 311–12.
62. JARVIS, A. W. (1974) A solid-phase radioimmunoassay for the detection of staphylococcal enterotoxins in dairy products. *New Zealand Journal of Dairy Science and Technology*, **9** (2), 37–9.
63. JOHNSON, H. M., BUKOVIC, J. A. & KAUFFMAN, P. E. (1972) Staphylococcal enterotoxins A and B: solid-phase radioimmunoassay in food. *Abstracts of the Annual Meeting of the American Society for Microbiology*, **72**, 23.
64. JOHNSON, H. M., BUKOVIC, J. A. & KAUFFMAN, P. E. (1973) Staphylococcal enterotoxins A and B: solid-phase radioimmunoassay in food. *Applied Microbiology*, **26** (3), 309–13.
65. JOHNSON, H. M., BUKOVIC, J. A., EISENBERG, W. V. & VAZQUEZ, A. W.

(1973) Antigenic properties of some insects involved in food contamination. *Journal of the Association of Official Analytical Chemists*, **56** (1), 63–5.
66. JOHNSTON, L. A. Y., TRACEY-PATTE, P., DONALDSON, R. A. & PARKINSON, B. (1982) A screening test to differentiate cattle meat from horse, donkey, kangaroo, pig and sheep meats. *Australian Veterinary Journal*, **59** (2), 59.
67. JOURDAN, P. S., MANSELL, R. L. & WEILER, E. W. (1982) Radioimmunoassay for the citrus bitter principle, naringin, and related flavonoid-7-O-neohesperidosides. *Planta Medica*, **44** (2), 82–6.
68. JOURDAN, P. S., WEILER, E. W. & MANSELL, R. L. (1983) Radioimmunoassay for naringin and related flavanone 7-neohesperidosides using a tritiated tracer. *Journal of Agricultural and Food Chemistry*, **31** (6), 1249–55.
69. KANG'ETHE, E. K., JONES, S. J. & PATTERSON, R. L. S. (1982) Identification of the species origin of fresh meat using an enzyme-linked immunosorbent assay procedure. *Meat Science*, **7** (3), 229–40.
70. KARISTO, T., TANHUANPAEAE, E., UUSI-RAUVA, A. & ANTILA, M. (1982) Progesterone concentration of milk and some dairy products. *Meijeritieteellinen Aikakauskirja*, **40** (2), 20–31.
71. KAUFFMAN, P. E. (1980) Enzyme immunoassay for staphylococcal enterotoxin A. *Journal of the Association of Official Analytical Chemists*, **63** (5), 1138–43.
72. KIMURA, L., JOYO, B., SHIRAKI, K., SASAKI, R. & HOKAMA, Y. (1980) Detection of ciguatoxin in fish tissues by radioimmunoassay (RIA). *Federation Proceedings*, **39** (3, II), 1141.
73. KIMURA, L. H., ABAD, M. A., YOKUCHI, L. A., HOKAMA, J. L. R. Y. & HOKAMA, Y. (1983) Solid phase immunoenzyme linked assay (ELISA) for the direct detection of Ciguatoxin (CTX) in fish muscle. *Federation Proceedings*, **42** (3), 555.
74. KNAPEN, F. VAN, FRAMSTAD, K. & RUITENBERG, E. J. (1976) Detection of *Trichinella spiralis* in pigs intended for slaughter using the enzyme-linked immunosorbent assay (ELISA). *Tijdschrift Voor Diergeneeskunde*, **101** (17), 952–6.
75. KNAPEN, F. VAN & RUITENBERG, E. J. (1979) Report 1977–1978 concerning *Trichinella spiralis* studies in The Netherlands. *Veterinary Quarterly*, **1** (3), 166–7.
76. KNAPEN, F. VAN., FRANCHIMONT, J. H. & LUGT, G. VAN DER (1982) Prevalence of antibodies to toxoplasma in farm animals in The Netherlands and its implication for meat inspection. *Veterinary Quarterly*, **4** (3), 101–5.
77. KOENIS, S. & MARTH, E. H. (1982) Behavior of *Staphylococcus aureus* in Cheddar cheese made with sodium chloride or a mixture of sodium chloride and potassium chloride. *Journal of Food Protection*, **45** (11), 996–1002.
78. KOPER, J. W., HAGENAARS, A. M. & NOTERMANS, S. (1980) Prevention of cross-reactions in the enzyme-linked immunosorbent assay (ELISA) for the detection of *Staphylococcus aureus* enterotoxin type B in culture filtrates and foods. *Journal of Food Safety*, **2** (1), 35–45.
79. KROES, R., HUIS IN'T VELD, L. G., SCHULLER, P. L. & STEPHANY, R. W. (1977) Methods for controlling the application of anabolics in farm animals.

In: *Environmental Quality and Safety Supplement*, Lo, F. C. & Rendel, J. (eds), Vol. 5, *Anabolic Agents in Animal Production*, FAO/WHO, pp. 192–202.

80. KUO, J. K. S. & SILVERMAN, G. J. (1980) Application of enzyme-linked immunosorbent assay for detection of staphylococcal enterotoxins in food. *Journal of Food Protection*, **43** (5), 404–7.
81. KURTH, L. & SHAW, F. D. (1983) Identification of the species origin of meat by electrophoretic and immunological methods. *Food Technology Australia*, **35**, 328–31.
82. LAM, N. C. S. I. (1981) Growth and enterotoxin production of *Staphylococcus aureus* (isolated from heat treated milk in a ten-month survey) in the manufacture of Cheddar cheese made from heat treated milk. *Dissertation Abstracts International B*, **42** (2), 551.
83. LAU, H. P., GAUR, P. K. & CHU, F. S. (1981)* Preparation and characterization of aflatoxin B_2 α-hemiglutarate and its use for the production of antibody against aflatoxin B_1. *Journal of Food Safety*, **3** (1), 1–13.
84. LEE, S. & CHU, F. S. (1981) Radioimmunoassay of T-2 toxin in corn and wheat. *Journal of the Association of Official Analytical Chemists*, **64** (1), 156–61.
85. LEE, S. & CHU, F. S. (1981) Radioimmunoassay of T-2 toxin in biological fluids. *Journal of the Association of Official Analytical Chemists*, **64** (3), 684–8.
86. LEFIER, D. & COLLIN, J. C. (1982) ELISA of bovine kappa-casein. *Lait*, **62** 541–8.
87. LEMBKE, J. & TEUBER, M. (1979) Detection of bacteriophages in whey by an enzyme-linked immunosorbent assay (ELISA). *Milchwissenschaft*, **34** (8), 457–8.
88. LEWIS, G. E. JR., KULINSKI, S. S., REICHARD, D. W. & METZGER, J. F. (1981) Detection of *Clostridium botulinum* type G toxin by enzyme-linked immunosorbent assay. *Applied and Environmental Microbiology*, **42** (6), 1018–22.
89. LIEMANN, F. & MUSCHKE, M. (1981) Radioimmunoassay for direct detection of diethylstilboestrol glucuronide and other oestrogenic stilbenes for routine residue control. *Archiv für Lebensmittelhygiene*, **32** (4), 110–15.
90. LINKROTH, S. & NISKANEN, A. (1977) Double antibody solid-phase radioimmunoassay for staphylococcal enterotoxin A. *European Journal of Applied Microbiology and Biotechnology*, **4** (2), 137–43.
91. LJUNGSTRÖM, I., ENGVALL, E. & RUITENBERG, E. J. (1974) ELISA — Enzyme-linked immunosorbent assay — in serological diagnosis of *Trichinella spiralis* infection. *Proceedings of Third International Congress of Parasitology*, Vols I, II, III, World Federation of Parasitologists, Congress Centre, Exhibition Grounds, Munich, Aug. 25–31, Facta Publications, Vienna, pp. 1204–5.

*Note: This paper has also been cited as 1980 probably because this date appears on the journal cover whereas 1981 appears on the title pages of all papers in this volume.

92. LUEHY, J. (1978) Review of methods for determination of aflatoxins B, G and M in foods and feed. *Mitteilungen aus dem Gebeite der Lebensmitteluntersuchung und Hygiene*, **69** (2), 200–19.
93. MALMFORS, B. & ANDRESEN, O. (1975) Relationship between boar taint intensity and concentration of 5-α-androst-16-en-one in boar peripheral plasma and back fat. *Acta Agriculturae Scandinavica*, **25** (2), 92–6.
94. MANSELL, R. L. & WEILER, E. W. (1980) Immunological tests for the evaluation of citrus quality. *ACS Symposium Series*, **143**, 341–59.
95. MANSELL, R. L. & WEILER, E. W. (1980) Radioimmunoassay for the determination of limonin in citrus. *Phytochemistry*, **19** (7), 1403–7.
96. MANSELL, R. L., MCINTOSH, C. A. & VEST, S. E. (1983) An analysis of the limonin and naringin content of grapefruit juice samples collected from Florida State test houses. *Journal of Agricultural and Food Chemistry*, **31**, 156–62.
97. MCINTOSH, C. A. & MANSELL, R. L. (1983) Distribution of limonin during the growth and development of leaves and branches of *Citrus paradisi*. *Journal of Agricultural and Food Chemistry*, **31** (2), 319–26.
98. MENZEL, E. J. & GLATZ, F. (1981) Radioimmunological determination of native and heat denatured soy protein. *Zeitschrift für Lebensmittel-untersuchung und Forschung*, **172** (1), 12–19.
99. METZGER, J. F. & JOHNSON, A. D. (1977) A modified radioimmunoassay for *Staphylococcus aureus* enterotoxins utilizing *S. aureus* Serotype 1. *Abstracts of the Annual Meeting of the American Society for Microbiology*, **77**, 259.
100. MILLER, B. A., REISER, R. F. & BERGDOLL, M. S. (1978) Detection of staphylococcal enterotoxins A, B, C, D and E in foods by radioimmunoassay, using staphylococcal cells containing protein A as immunoadsorbent. *Applied and Environmental Microbiology*, **36** (3), 421–6.
101. MINNICH, S. A., HARTMAN, P. A. & HEIMSCH, R. C. (1982) Enzyme immunoassay for detection of salmonellae in foods. *Applied and Environmental Microbiology*, **43** (4), 877–83.
102. MIURO, T., KOUNO, H. & KITAGAWA, T. (1981) Detection of residual penicillin in milk by sensitive EIA. *Journal of Pharmacobio-Dynamics*, **4** (9), 706–10.
103. MIYAHARA, J., SHIRAKI, K., AKAU, R., CHUNG, R., JOYO, B., KIMURA, L. H. & HOKAMA, Y. (1980) Comparative examination of the radioimmunoassay (RIA) for detection of ciguatoxin in fish tissue and the pharmacological effect of extracted ciguatoxin on mammalian atria. *Federation Proceedings*, **39** (3, II), 1141.
104. MORGAN, M. R. A., MCNERNEY, R., MATTHEW, J. A., COXON, D. T. & CHAN, H. W-S. (1983) An enzyme-linked immunosorbent assay for total glycoalkaloids in potato tubers. *Journal of the Science of Food and Agriculture*, **34**, 593–8.
105. MORGAN, M. R. A., MATTHEW, J. A., MCNERNEY, R. & CHAN, H. W-S. (1982) The immunoassay of Ochratoxin A. *Proceedings of the Fifth International IUPAC Symposium on Mycotoxins and Phycotoxins*, Vienna, pp. 32–5.

106. MORGAN, M. R. A., MCNERNEY, R. & CHAN, H. W-S. (1983) The enzyme-linked immunosorbent assay of ochratoxin A in barley. *Journal of the Association of Official Analytical Chemists*, **66** (6), 1481–4.
107. MUSCHKE, M. & LIEMANN, F. (1979) Potential application of radioimmunoassay to food analysis, using diethylstilboestrol as an example. *Lebensmittelchemie und Gerichtliche Chemie*, **33** (2), 41.
108. NERENBERG, C., TSINA, I. & MATIN, S. (1982) Radioimmunoassay of oxfendazole in sheep fat. *Journal of the Association of Official Analytical Chemists*, **65** (3), 635–9.
109. NEWSOME, W. H. & SHIELDS, J. B. (1981) A radioimmunoassay for benomyl and methyl 2-benzimidazolecarbamate on food crops. *Journal of Agricultural and Food Chemistry*, **29** (2), 220–2.
110. NISKANEN, A. (1977) Staphylococcal enterotoxins and food poisoning. Production, properties and detection of enterotoxins. *Materials and Processing Technology*, No. 19, Technical Research Centre of Finland.
111. NIYOMVIT, N., STEVENSON, K. E. & MCFEETERS, R. F. (1978) Detection of staphylococcal enterotoxin B by affinity radioimmunoassay. *Journal of Food Science*, **43** (3), 735–9.
112. NOLETO, A. L. & BERGDOLL, M. S. (1982) Production of enterotoxin by a *Staphylococcus aureus* strain that produces three identifiable enterotoxins. *Journal of Food Protection*, **45** (12), 1096–7.
113. NOTERMANS, S., BOOT, R., TIPS, P. D., NOOIJ, M. P. DE (1983) Extraction of staphylococcal enterotoxins (SE) from minced meat and subsequent detection of SE with enzyme-linked immunosorbent assay (ELISA). *Journal of Food Protection*, **46** (3), 238–41, 244.
114. NUTI, L. C., WENTWORTH, B. C., KARAVOLAS, H. J., TYLER, W. J. & GINTHER, O. J. (1975) Comparison of radioimmunoassay and gas-liquid chromatography analyses of progesterone concentrations in cow's milk. *Proceedings of the Society for Experimental Biology and Medicine*, **149** (4), 877–80.
115. OLSVIK, O., GRANUM, P. E. & BERDAL, B. P. (1982) Detection of *Clostridium perfringens Type A* enterotoxin by ELISA. *Acta Pathologica, Microbiologica et Immunologica Scandinavica, B* (*Microbiology*), **90** (6), 445–7.
116. OLSVIK, O., MYHRE, S., BERDAL, B. P. & FOSSUM, K. (1982) Detection of staphylococcal enterotoxin A, B and C in milk by an ELISA procedure. *Acta Veterinaria Scandinavica*, **23** (2), 204–10.
117. ORTH, D. S. (1977) Statistical analysis and quality control in radioimmunoassays for staphylococcal enterotoxins A, B, and C. *Applied and Environmental Microbiology*, **34** (6), 710–14.
118. OUDERAA, F. VAN DER & HAAS, A. (1981) Use of immunoassays to detect enterotoxin B of *Staphylococcus aureus* in foods. *Antonie van Leeuwenhoek*, **47**, 186–7.
119. PATTERSON, M. R., SPENCER, T. L. & WHITTAKER, R. G. (1983) Enzyme linked immunosorbent assays for speciating meat. In: *Horizon 90: How to survive the 80s. 16th Annual Convention Australian Institute of Food Science and Technology*, p. 16.
120. PESTKA, J. J., GUAR, P. K. & CHU, F. S. (1980) Quantitation of aflatoxin B_1

and aflatoxin B_1 antibody by an enzyme-linked immunosorbent microassay. *Applied and Environmental Microbiology*, **40** (6), 1027–31.

121. PESTKA, J. J., LI, Y., HARDER, W. O. & CHU, F. S. (1981) Comparison of radioimmunoassay and enzyme-linked immunosorbent assay for determining aflatoxin M_1 in milk. *Journal of the Association of Official Analytical Chemists*, **64** (2), 294–301.
122. PESTKA, J. J., LEE, S. C., LAU, H. P. & CHU, F. S. (1981) Enzyme-linked immunosorbent assay for T-2 toxin. *Journal of the American Oil Chemists Society*, **58** (12), 940A–4A.
123. PESTKA, J. J., STEINERT, B. W. & CHU, F. S. (1981) Enzyme linked immunosorbent assay for detection of Ochratoxin A. *Applied and Environmental Microbiology*, **41**, 1472–4.
124. POBER, Z. & SILVERMAN, G. J. (1976) A modified RIA determination for staphylococcal enterotoxin B in food. *Abstracts of the Annual Meeting of the American Society for Microbiology*, **76**, 188.
125. POBER, Z. & SILVERMAN, G. J. (1977) Modified radioimmunoassay determination for staphylococcal enterotoxin B in foods. *Applied and Environmental Microbiology*, **33** (3), 620–5.
126. PORATH, J. (1970) Radioimmunoassays as a means of determining small quantities of specific proteins. In: *Evaluation of Novel Protein Products*, Bender, A. E., Kihlberg, R., Loefqvuist, B. and Munck, L. (eds), Pergamon Press, Oxford, pp. 285–6.
127. RAUCH, P., FUKAL, L. & KAS, J. (1982) Determination of subresidual proteolytic activities in foods. In: *Recent Developments in Food Analysis*, Baltes, W. (ed.), conference proceedings, European Federation of Chemical Societies, Verlag Chemie, Weinheim, pp. 240–6.
128. REICHERT, C. A. & FUNG, D. Y. C. (1976) Thermal inactivation and subsequent reactivation of staphylococcal enterotoxin B in selected liquid foods. *Journal of Milk and Food Technology*, **39** (8), 516–20.
129. ROBERN, H., DIGHTON, M., YANO, Y. & DICKIE, N. (1975) Double antibody radioimmunoassay for staphylococcal enterotoxin C2. *Applied Microbiology*, **30** (4), 525–9.
130. ROBERN, H., GLEESON, T. M. & SZABO, R. A. (1978) Double antibody radioimmunoassay for staphylococcal enterotoxins A and B. *Canadian Journal of Microbiology*, **24** (4), 436–9.
131. ROBERN, H. & GLEESON, T. M. (1978) The use of polyethylene glycol in radioimmunoassay of staphylococcal enterotoxins. *Canadian Journal of Microbiology*, **24** (6), 765–6.
132. ROBINS, R. J., MORGAN, M. R. A., RHODES, M. J. C. & FURZE, J. M. (1984) An enzyme-linked immunosorbent assay for quassin and closely related metabolites. *Analytical Biochemistry*, **136**, 145–56.
133. ROBISON, B. J., PRETZMAN, C. I. & MATTINGLY, J. A. (1983) Enzyme immunoassay in which a myeloma protein is used for detection of salmonellae. *Applied and Environmental Microbiology*, **45** (6), 1816–21.
134. ROGDAKIS, E., ENSINGER, U. & FABER, H. VON (1979) Concentration of cAMP in the *Longissimus dorsi* muscle of Pietrain and Edelschwein Large White) swine. *Zühtungskunde*, **51** (1), 48–51.
135. RUITENBERG, E. J., STEERENBERG, P. A., BROSI, B. J. M., BUYS, J.,

LJUNGSTRÖM, I. & ENGVALL, E. (1974) Application of ELISA for the serodiagnosis of *T. spiralis* infections in pigs under slaughterhouse conditions. *Proceedings of Third International Congress of Parasitology*, Vols I, II, III, World Federation of Parasitologists, Congress Centre, Exhibition Grounds, Munich, Aug. 25–31, Facta Publications, Vienna.

136. RUITENBERG, E. J., STEERENBERG, P. A., BROSI, B. J. M. & BUYS, J. (1976). Reliability of the enzyme-linked immunosorbent assay (ELISA) for the serodiagnosis of *Trichinella spiralis* infections in conventionally raised pigs. *Tijdschrift voor Diergeneeskunde*, **101** (2), 57–70.
137. SAUNDERS, G. C. & BARTLETT, M. L. (1977) Double-antibody solid-phase enzyme immunoassay for the detection of staphylococcal enterotoxin A. *Applied and Environmental Microbiology*, **34** (5), 518–22.
138. SCHOPPER, D. (1983) Detecting the synthetic anabolic substance trenbolone acetate in slaughter animals. *Fleischwirtschaft*, **63** (3), 406–7.
139. SCHWEIZER, A., HANNIG, K., GUNTHER, H. O. & BAUDNER, S. (1982) Immunological detection of meat from turkey. In: *Recent Developments in Food Analysis*, Baltes, W. (ed.) conference proceedings, European Federation of Chemical Societies, Verlag Chemie, Weinheim, pp. 449–52.
140. SIMON, E. & TERPLAN, G. (1977) Detection of staphylococcal enterotoxin B by means of the ELISA-test. *Zentralblatt für Veterinärmedizin B*, **24** (10), 842–4.
141. SUN, P. S. & CHU, F. S. (1977) A simple solid-phase radioimmunoassay for aflatoxin B_1. *Journal of Food Safety*, **1** (1), 67–75.
142. SWAMINATHAN, B. & AYES, J. C. (1980) A direct immunoenzyme method for the detection of Salmonellae in foods. *Journal of Food Science*, **45**, 352–5, 361.
143. VALLEJO, R. P. & ERCEGOVICH, C. D. (1979) Analysis of potato for glycoalkaloid content by radioimmunoassay (RIA). *Trace Organic Analysis: A New Frontier in Analytical Chemistry, Proceedings of the Ninth Materials Research Symposium*, National Bureau of Standards, USA, Special Publication 519, pp. 333–40.
144. VALLEJO, R. P., BOGUS, E. R. & MUMMA, R. O. (1982) Effect of hapten structure and bridging groups on antisera specificity in parathion immunoassay development. *Journal of Agricultural and Food Chemistry*, **30**, 572–80.
145. VOGT, K. (1980) Simplified extraction and purification procedure for radioimmunological assay of diethylstilboestrol in meat, liver and kidney. *Archiv für Lebensmittelhygiene*, **31** (4), 138–41.
146. WAL, J. -M., BORIES, G. F., MAMAS, S. & DRAY, F. (1975) Radioimmunoassay of penicilloyl groups in biological fluids (including milk). *FEBS Letters*, **57** (1), 9–13.
147. WAL, J. -M. & BORIES, G. F. (1981) *In vitro* penicillin aminolysis: application to a radioimmunoassay of trace amounts of penicillin. *Analytical Biochemistry*, **114** (2), 263–7.
148. WALSH, J. H. (1980) Estimation of the pantothenic acid content of foods using a microbiological assay and a radioimmunoassay. *Dissertation Abstracts International B*, **41** (3), 900.

149. Walsh, J. H., Wyse, B. W. & Hansen, R. G. (1980) A comparison of microbiological and radioimmunoassay methods for the determination of pantothenic acid in foods. *Journal of Food Biochemistry*, **3** (4), 175–89.
150. Walsh, J. H., Wyse, B. W. & Hansen, R. G. (1981) Pantothenic acid content of 75 processed and cooked foods. *Journal of the American Dietetic Association*, **78** (2), 140–4.
151. Weiler, E. W. & Mansell, R. L. (1980) Radioimmunoassay of limonin using a tritiated tracer. *Journal of Agricultural and Food Chemistry*, **28** (3), 543–5.
152. Whittaker, R. G., Spencer, T. L. & Copland, J. W. (1982) Enzyme-linked immunosorbent assay for meat species testing. *Australian Veterinary Journal*, **59** (4), 125.
153. Whittaker, R. G., Spencer, T. L. & Copland, J. W. (1983) An enzyme-linked immunosorbent assay for species identification of raw meat. *Journal of the Science of Food and Agriculture*, **34**, 1143–8.
154. Wie, S. I. & Hammock, B. D. (1982) The use of enzyme-linked immunosorbent assays for the determination of Triton X nonionic detergents. *Analytical Biochemistry*, **125**, 168–76.
155. Wie, S. I., Sylvester, A. P., Wing, K. D. & Hammock, B. D. (1982) Synthesis of haptens and potential radioligands and development of antibodies to insect growth regulators of diflubenzuron and BAY SIR 8514. *Journal of Agricultural and Food Chemistry*, **30**, 943–8.
156. Wie, S. I. & Hancock, B. D. (1982) Development of enzyme-linked immunosorbent assays for residue analysis of diflubenzuron and BAY SIR 8514. *Journal of Agricultural and Food Chemistry*, **30**, 949–57.
157. Wikman-Coffelt, J. & Berg, H. W. (1976) Radioimmunoassay method for analysis of ethylcarbamate in wine. *American Journal of Enology and Viticulture*, **27** (3), 115–17.
158. Wing, K. D. & Hammock, B. D. (1980) Immunochemical methods to detect pesticide residues. *California Agriculture*, **34** (3), 34–5.
159. Wolford, S. T. & Argoudelis, C. J. (1979) Measurement of estrogens in cow's milk, human milk, and dairy products. *Journal of Dairy Science*, **62** (9), 1458–63.
160. World Health Organization (1976) The enzyme-linked immunosorbent assay (ELISA) and its potential for food hygiene. In: *Microbiological Aspects of Food Hygiene*, Report of a WHO Expert Committee with the participation of FAO, WHO Technical Report Series, no. 598.

INDEX TO BIBLIOGRAPHY

Numbers after each entry correspond to the reference numbers in the Bibliography

Aflatoxin, 12, 13, 35, 83, 92, 120, 121, 141
Amatoxins, 38
Antibiotics, 102, 146, 147

Bacteriophage, 87
Beer, 34
Bitter principles, 67, 68, 94–7, 132, 151
Boar taint, 16, 23–5, 43, 93

Caffeine, 31
Cereals, 84, 92, 106
Ciguatoxin, 10, 72, 73, 103
Clostridium toxins, 88, 115

Detergents, 154

Enzymes, 34, 61, 127

Fat, 35
Fish, 10, 17, 72, 73, 103
Fruit and fruit products, 40, 61, 67, 68, 94–7, 151

Glycoalkaloids, 28, 104, 143

Hygromycins, 42

Insects, 63

Legumes, 41, 55, 98, 127

Meat, 3, 11, 15, 16, 23, 37, 43, 46–9, 54–7, 66, 69, 76, 79, 87, 91, 93, 108, 113, 119, 134–6, 139, 145, 152, 153
Milk and milk products, 5, 20, 22, 45, 52, 59, 62, 70, 77, 82, 86, 87, 92, 102, 114, 116, 121, 146, 159
Mushrooms, 26, 38

Ochratoxins, 1, 105, 106, 123

Pesticides, 4, 36, 109, 144, 155–8

Review paper, 30, 52, 75, 94, 107, 160

Salmonella, 101, 123, 142
Soft drinks, 31, 40
Species testing, 3, 66, 69, 87, 119, 152, 153
Staphylococcus aureus toxins, 2, 6, 8, 9, 20, 21, 27, 32, 33, 39, 44, 62–4, 71, 77, 78, 80, 82, 90, 99, 100, 110–13, 116–18, 124, 125, 128–31, 137, 140
Steroids, 7, 15–19, 22–5, 43, 45–9, 54, 56–8, 70, 79, 89, 93, 107, 114, 138, 145, 159
Sweetness, 60

Thaumatin, 60
Trichinella, 11, 74, 75, 91, 135, 136
Tricothecins, 84, 85, 122

Vegetables, 14, 25, 28, 50, 61, 104, 143
Virus, 14, 50
Vitamins, 59, 148–50

2
Principles of Immunoassay

B. A. MORRIS

Division of Clinical Biochemistry, Department of Biochemistry, University of Surrey, Guildford, UK

INTRODUCTION

The technique of radioimmunoassay was developed quite accidentally by Yalow and Berson in the late 1950s (Yalow & Berson, 1959; 1960), as they endeavoured to separate ^{131}I-labelled insulin that had been metabolised *in vivo* from that which had remained intact. From those humble beginnings, the technique has become one of the most sensitive analytical methods available although, in food analysis, this sensitivity is not always required. It is, in addition, an extremely simple and specific technique. Today, immunoassay covers not only the classical immunoassays, as conceived by Berson and Yalow, but also labelled-antibody techniques, such as immunometric assays and the enzyme-linked immunosorbent assays (ELISAs), as well as many other variations. In fact, immunoassay has come to mean different things to different people. The true immunoassays are characterised by the fact that they use a labelled form of the antigen or antibody for quantification and a non-precipitating antibody in the primary reaction. These features distinguish them from other quantitative immunological techniques, such as Laurell rocket and Mancini immunoprecipitation techniques, where no labelled reactant is involved and a precipitating antibody is used. One advantage, therefore, of immunoassay is the very high sample throughput that can be achieved because one does not have to wait for an antigen–antibody precipitate to form, as one does with immunoprecipitation techniques. The interaction between antigen and antibody in the liquid phase can be very fast and this results in a very rapid assay.

This chapter is predominantly concerned with the classical immunoassay developed by Berson and Yalow, although the labelled-antibody technique will also be briefly considered. Enzyme-linked immunosorbent assay, which probably will play an important part in food analysis in the future, is the subject of Chapter 3.

CLASSICAL IMMUNOASSAY

This is the original immunoassay system, as devised by Berson & Yalow. It is characterised by the fact that it is a limited reagent assay; in other words, there is less binding protein, antibody, present than there is antigen, and in order to quantitate the system, one has to use a labelled form of the antigen being measured.

This labelled antigen is allowed to react with a fixed, but limited, number of specific antibodies in such relative concentrations that only about 50% of the added label is bound by the antibody fraction (Fig. 1). At the end of the reaction period, the portion of the labelled antigen which has bound to antibody is separated from that which has not, in a process known as phase separation. The concentration of the label in either the antibody-bound or free fraction is then determined.

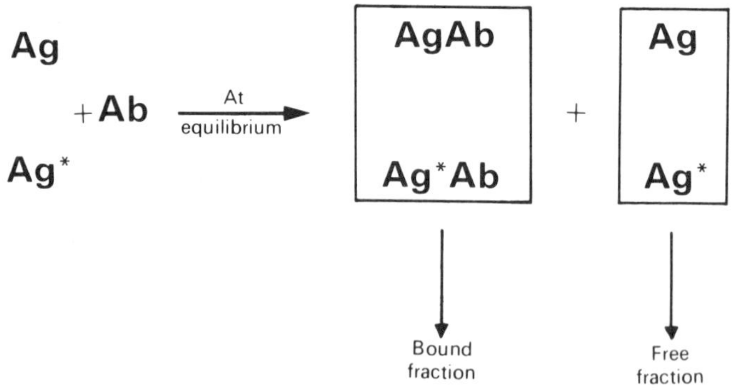

Fig. 1. Principle of immunoassay: Ag, unlabelled antigen; Ag*, labelled antigen; Ab, antibody; AgAb, antigen–antibody complex.

If a quantity of unlabelled antigen, the analyte, present in either a standard solution or sample, is simultaneously incubated with this mixture of labelled antigen and antibody, the unlabelled antigen will compete with the labelled antigen for the limited number of binding sites on the antibodies present. In other words, the more unlabelled antigen molecules that are present, the fewer labelled molecules will be bound by antibody. If the label and antibody are incubated with known amounts of antigen, a standard curve can then be derived (Fig. 2), and the concentration of antigen in samples determined by interpolation.

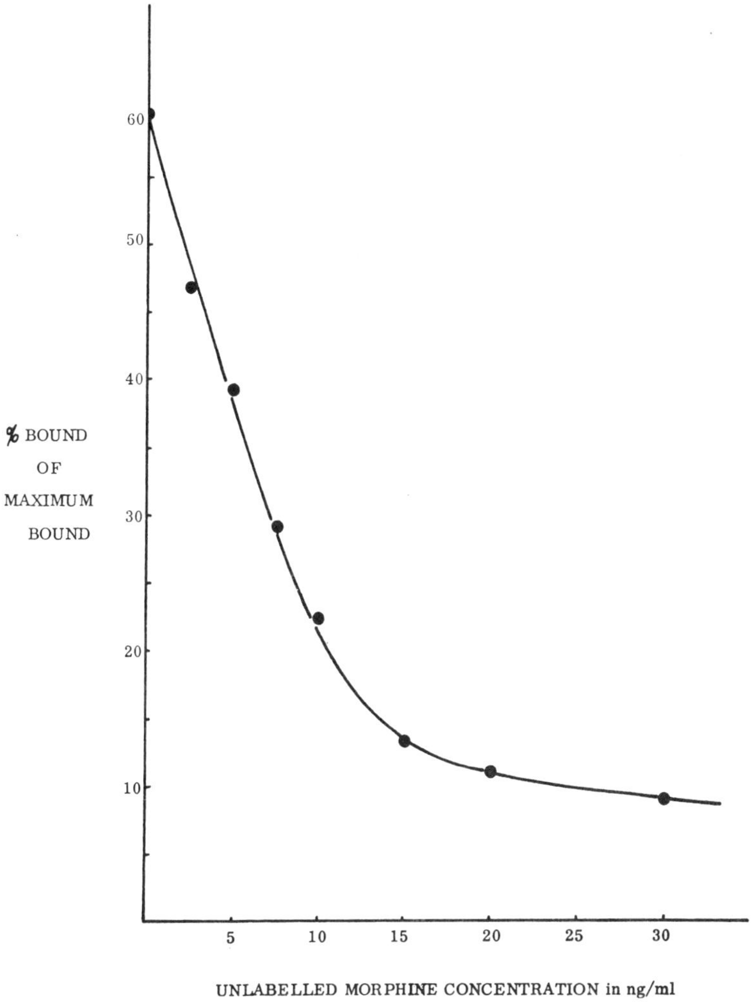

Fig. 2. A typical immunoassay standard curve. Antiserum: anti-6-succinyl-morphine, G/G/1-VIIA, 1:60. Label: 1 pM ^{3}H-dihydromorphine/tube.

It therefore follows that one requires four basic reagents to perform a classical immunoassay:

(a) a standard preparation of the antigen
(b) a labelled form of the antigen
(c) an antiserum specific for the antigen
(d) a phase separation system

Many immunoassayists would add as a fifth requirement to that list a good quality control system (Jeffcoate, 1981). This is particularly important in immunoassay, since one is dealing with biological reagents. Unlike chemical reagents, which are relatively consistent in composition and behaviour from one laboratory to another, the reactivity of antibodies and antigens towards each other is influenced by the environment in which the reaction takes place. In addition, each batch of antiserum is a unique reagent, having its own particular specificity for the antigen in question. Therefore, a good quality control system in immunoassay is probably more important than in other techniques.

LABELLED ANTIGENS

A number of different molecules and compounds have been used to label antigens in order to quantitate the immunoassay system and these are shown in Table 1.

The first labels used were the radioisotopes, ^{131}I and then ^{125}I, which were used to label the protein and polypeptide hormones. As the technique was extended to the measurement of steroids and drugs, which could not be iodinated directly, tritium- (^{3}H) and ^{14}C-labelled antigens were used in their place. However, not only are the immunoassays using these radionucleides, especially those using ^{14}C, inherently less sensitive than those using the radioisotopes of iodine, but they also suffer from the fact that both ^{3}H and ^{14}C are β-emitters. This means that, in order to determine their activity, they must first be incorporated into scintillation cocktails, and this in turn involves quenching and chemiluminescent problems during the counting process. In addition, the throughput is slow, since it takes 4–5 min to count each sample in the case of tritium. The iodine radionucleides, on the other hand, are γ-emitters, and can be counted directly in a γ-counter without any sample preparation. The assay tubes, containing the precipitated fraction, can be

Table 1
Types of Label Used in Immunoassay

1. Radioisotopes:
(a) ^{125}I and ^{131}I
(b) ^{3}H
(c) ^{14}C
2. Enzymes
3. Fluorescent molecules
4. Luminescent molecules:
(a) Chemiluminescent
(b) Bioluminescent
5. Spin radicals
6. Red blood cells
7. Phages

put straight into the counter and the amount of activity determined in 15–30 s.

Antigens labelled with tritium and ^{14}C require, for their preparation, the facilities usually only available in the factories of the commercial radiopharmaceutical manufacturers. However, if a grade C radioisotope laboratory is available, one can produce one's own ^{125}I-labelled antigens, providing that the molecule one wishes to label possesses either a tyrosine or histidine residue.

There are a number of methods available for introducing radioiodine into a molecule. They all involve the oxidation of $Na^{125}I$ to yield nascent iodine. The most widely used method is that of Greenwood *et al.* (1963) involving chloramine-T, which yields hypochlorous acid in aqueous solution. However, the oxidation–reduction conditions of this method destroy the immunoreactivity of some molecules and it is necessary to resort to milder oxidation conditions when labelling them. The mildest of these is achieved with the lactoperoxidase method of Marchalonis (1969) modified by Thorell & Johansson (1971). More recently, a simpler method of radioiodination has been reported using a solid-phase catalyst, iodogen (Fraker & Speck 1978; Salacinski *et al.*, 1979). Whilst considerably easier to perform than the chloramine-T method, it is not as effective as lactoperoxidase with the more labile antigens.

If the molecule does not possess either a tyrosine or histidine residue, then all is not lost, since one can resort to the use of radioiodination tags: these are small molecules whose structures mimic either tyrosine or histidine. They can be attached by covalent linkage to the antigen, either before or after radioiodination, to produce a labelled antigen for use in an immunoassay. If the antigen is a hapten, one must be careful to avoid using the same conjugation method and/or the same point of attachment on the molecule as was used to prepare the immunogen. If one does use the same conjugation method and position on the molecule to attach the tag, the chances are that the antibody will recognise the labelled antigen much more strongly and with a higher energy of reaction than the antigen itself. This phenomenon is known as 'bridge recognition' (Gilby & Jeffcoate 1973; Robinson *et al.*, 1975*a*) in which the antibody recognises not only the antigen, but also the bridge linking it to the tag. To avoid this problem, one should either conjugate the tag to the molecule at a different position or use a different conjugation method. This problem also occurs in the preparation of enzyme-labelled haptens, see Chapter 3.

Sometimes the situation arises that either the facilities for dealing with radioactivity are not available, or the environment in which one works precludes the use of radioactivity. This problem can be overcome by using one of the non-isotopic labels listed in Table 1. The principal ones are the enzymes and the fluorescent and luminescent molecules, and these are the subjects of Chapters 3 and 4.

ANTIBODIES FOR IMMUNOASSAY

Theoretical Considerations

In the introduction the importance of avid antibodies in immunoassay was stressed, and this is one of the features which distinguishes classical immunoassay from other immunological quantitation techniques. Why the emphasis on avidity?

Because the antigen–antibody complex in immunoassay does not form a precipitate, the reaction obeys the Law of Mass Action, with the result that not only can the antigen combine with its antibody to form the complex, but the complex can also dissociate:

$$\mathrm{Ag} + \mathrm{Ab} \rightleftharpoons \mathrm{AgAb}$$

At equilibrium, the complex is dissociating as fast as it is being formed. Therefore, the amount of complex present is really a measure of the energy of reaction, i.e. avidity, of the antibodies being used. The higher the avidity, the greater the amount of antigen–antibody complex formed.

The avidity of an antibody in immunoassay is important for three reasons. It affects:

(a) the detection limit or sensitivity of the assay
(b) the association time, i.e. incubation time, of the assay
(c) the phase separation system used

To appreciate how the detection limit of an immunoassay is determined by the avidity of the antibody being used, let us reconsider the Law of Mass Action equation. Suppose that labelled antigen molecules are incubated with a limited number of infinitely high avid antibodies. At equilibrium the reaction will be shifted far over to the right, with the production of antigen–antibody complex and little or no free antibody. If one now introduces into that system extra antigen, since the reaction is already shifted to the right, the extra number of antigen molecules will have very little effect on the total amount of antigen–antibody complex formed. Therefore, if one allows the two forms of antigen to compete simultaneously for the antibody binding sites, the amount of complex will remain constant, independent of the amount of antigen present in the system, although the more antigen there is present, the less labelled antigen will be bound by antibody. If one derives a standard curve with this type of antibody, then providing the conditions have been optimised, one will get a very rapid rate of fall in the percentage of labelled antigen bound as one increases the concentration of unlabelled antigen. This very sensitive standard curve will result in an assay with a low detection limit, providing that the precision of the replicates is good.

Now in contrast, let us consider an immunoassay using a low avid antibody. If one incubates labelled antigen molecules with a limited number of these antibodies, then at equilibrium, the reaction will not be shifted so far over to the right as it was with the infinitely high avid antibodies. What is more important is that at equilibrium, besides the antigen–antibody complex and free labelled antigen, one will also have *free antibody* present, despite the fact that an antibody concentration was chosen which bound say 50% of the labelled molecules present. If, into that system one now adds extra antigen, in the form of unlabelled antigen, the reaction will be shifted still further over to the right in accordance with the Law of Mass Action, and more antigen–antibody complex will be formed. Therefore, if a standard curve is derived with this type of antibody, then as the unlabelled antigen concentration is increased, the rate of fall in percent binding of the labelled antigen will not be so great. A shallow curve will result and the assay will thus not have as low a detection limit as that using the infinitely high avid antibody.

The other two aspects which are affected by the avidity of the antibody are incubation time and choice of phase separation system. The higher the avidity of an antibody, the faster equilibrium will be attained, and as a result, the shorter will be the incubation period required to achieve this. It is partly by selecting antisera of very high avidity — 'high kinetic energy' — that kit manufacturers have been able to reduce their assay incubation times.

Antibody Production

The production of an antiserum containing antibodies suitable for use in an immunoassay can be divided into three stages:

Preparation of immunogen
Immunisation and blood collection
Assessment of antisera

With regard to immunisation schedules, there are probably as many different ones as there are papers reporting them. Some authorities recommend giving monthly boosters after the priming injection, whilst others recommend either shorter or longer time intervals. Part of the explanation for this diversity of recommended methods is that there is very little objective information on this subject, due to the tremendous variation in immune response that occurs between individual animals, even between animals of the same group. This, and the cost of performing trials large enough to provide statistically valid comparisons, accounts for the lack of objective data and the very subjective recommendations which appear in the literature. Hurn (1971) extolled the benefits of leaving the primed animals for as long as possible before boosting them. But how long is 'as long as possible'?

I firmly believe that the only individual able to answer this question is the animal itself, by the way in which it responds to the priming injection. This approach requires careful monitoring of the specific antibody levels at regular intervals after the first injection. One usually sees an initial rise in circulating level of specific antibody, which eventually peaks, and then gradually declines. I believe that one should not boost the animal until the concentration of circulating specific antibody has *fallen* to a steady level. This may take weeks, or even months (Fig. 3).

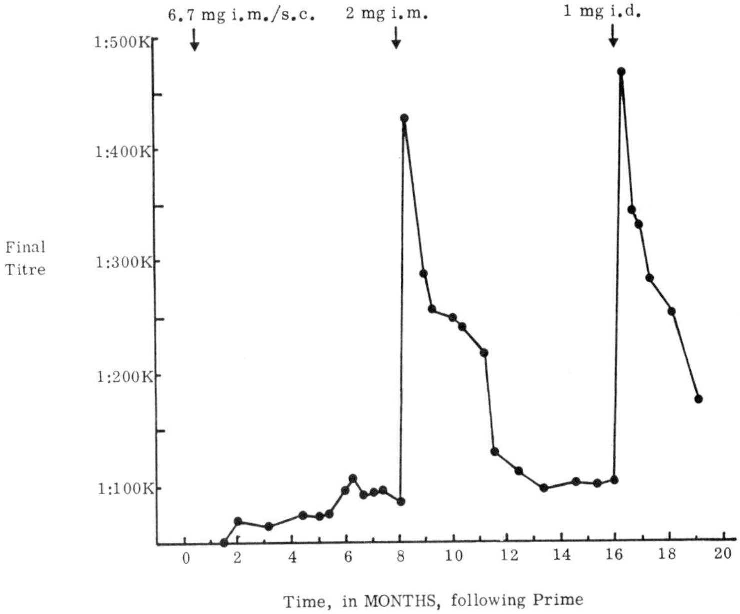

Fig. 3. Pattern of immune response in sheep **BI** immunised with 17β-oestradiol-6-carboxymethyloxime-ovalbumen. Route of injection of immunogen: i.m., intramuscular; s.c., subcutaneous; i.d., intradermal.

If one then gives a booster injection of approximately 40% of the amount of the material used for the priming dose, one usually sees a dramatic rise in antibody levels. This response peaks between 8 and 10 days after the booster injection, but the level then falls very rapidly afterwards, making it necessary to harvest as much blood as possible during the period whilst the levels are high. One can normally expect to obtain an immunological response to the first booster injection which is at least tenfold higher than that produced to the priming injection, if the latter was given by either the intramuscular or subcutaneous route. It is then necessary to continue to monitor the circulating antibody levels at regular intervals in order to determine when to give the second booster injection. This is given when the antibody levels have fallen again to a steady value.

Figure 3 shows the immune response in a sheep injected with an oestradiol–ovalbumen conjugate. Note that a period of 8 months was allowed to elapse between the priming injection and the first booster. This was not an arbitrarily decided interval, but dictated solely by the response of the animal in question. Sometimes, the time interval is shorter, sometimes it is longer; recently, we left a sheep producing a steroid antiserum for 15 months before boosting it. Admittedly that was a very long time, but patience was more than rewarded with the production of an extremely avid antiserum with a titre in excess of 1:2 000 000.

What is the rationale for leaving an animal for this length of time before boosting it? One first has to consider why the antibody levels fall after reaching a peak, and do not, on the other hand, continue to rise indefinitely, particularly after a booster injection. The elimination of the injected material from the depot at the injection site, thus reducing its immunogenic stimulus, may be part of the cause. However, the more likely explanation is to be found at the cellular level, where a safety mechanism exists to ensure that the concentration of circulating antibodies does not become so great that the increased viscosity of the blood impairs cardiac function. This state of affairs occurs in patients suffering from myeloma, where the body has lost its ability to contain antibody levels within reasonable limits.

In the normal individual, antibody levels are controlled by the suppressor and cytotoxic T-cells, which proliferate after the immune system has been stimulated and circulating antibody levels have increased. Once the production of antibody has been brought under control, the suppressor and cytotoxic T-cells return to their quiescent state and their influence on the remaining cells producing antibody is removed. The level of circulating antibody will then remain relatively constant. If the animal is now boosted, the antibody producing cells can multiply unhampered until they are again brought under control by the suppressor and cytotoxic T-cells. Therefore, in my opinion, there is no point in boosting an animal against a background of proliferating suppressor and cytotoxic T-cells, as is indicated by falling levels of circulating antibody from a peak value. One only secures the maximum response to a boost when these cells are in a quiescent phase.

Turning now to the species of animal used to produce antisera for immunoassay, there is now a wide choice available, depending on the nature of the immunogen and the use to which the antiserum is to be put. Rats can be used for screening different preparations prior to injecting the most immunogenic into a larger species (Robinson *et al.*, 1975*b*). Guinea pigs are not as widely used as they used to be, because they have to be bled by cardiac puncture under anaesthesia. They are, however, the species of choice for producing anti-insulin antibodies. Antisera to most large molecular weight compounds can be produced in rabbits, which are relatively cheap to keep and easy to maintain. However, the relatively small volume of antisera they produce per bleed, compared to the larger species, may be a limitation if one is looking for continuity of reagents, particularly for use by more than one laboratory. This problem may be overcome by raising the antisera in sheep or goats, which, additionally, are able to recognise small haptens much better than rabbits, thus producing higher titre antisera. If one requires second antibodies, particularly for use in liquid phase separation systems, then these are best raised in donkeys.

Assessment of Antisera

In immunoassay, antisera are assessed much more stringently than they are for other quantitative immunological techniques. An assessment of their usefulness is made in terms of their titre, i.e. the concentration of antibody present, their avidity, and finally their specificity. This last criterion is especially important in the case of antisera to small haptens, since it is possible to control unwanted cross-reactions by judicious selection of the point of attachment of the hapten to the carrier protein during the preparation of the immunogen (see Chapter 15).

Titre

One of the pitfalls that people fall into when selecting an antiserum from a number which are available for a particular assay is that they titre all of them and then select the ones with the highest titres for further evaluation. That is *not* the way to select the most appropriate bleed, because one may be missing an antiserum of exceptionally high avidity which may not be of quite such high titre as the others. So the emphasis should be on avidity rather than titre, i.e. quality rather than quantity. This is because not only will the avidity of the antiserum one chooses determine the detection limit of the assay being developed, but the higher the avidity of the antiserum, the shorter the assay incubation time. In addition, the choice of phase separation, too, will be controlled to a certain extent, in the case of small molecules, by the avidity of the antiserum being used. However, before one can select for avidity, one has to know the concentration of antibody present.

The titre of an antiserum is determined by means of an antiserum dilution curve, and is arbitrarily defined as that dilution (preferably final) of the antiserum which binds 50% of the immunoreactive portion of a specified quantity of the labelled antigen. Increasing dilutions of the antiserum under study are incubated with a constant amount of labelled antigen. At the end of the incubation period, the label which has bound to antibody is separated from that which has not and the amount associated with the bound fraction determined. After correction for the label which has bound non-specifically to the walls of the assay tube, the amount of label bound to antibody at each antiserum dilution is expressed as a percentage of the total quantity of label added to the tube. The percentage of label bound to antibody is plotted against the final dilution of the antiserum (Fig. 4). When the points are joined with a smooth line, a sigmoid curve results, showing the highest binding with

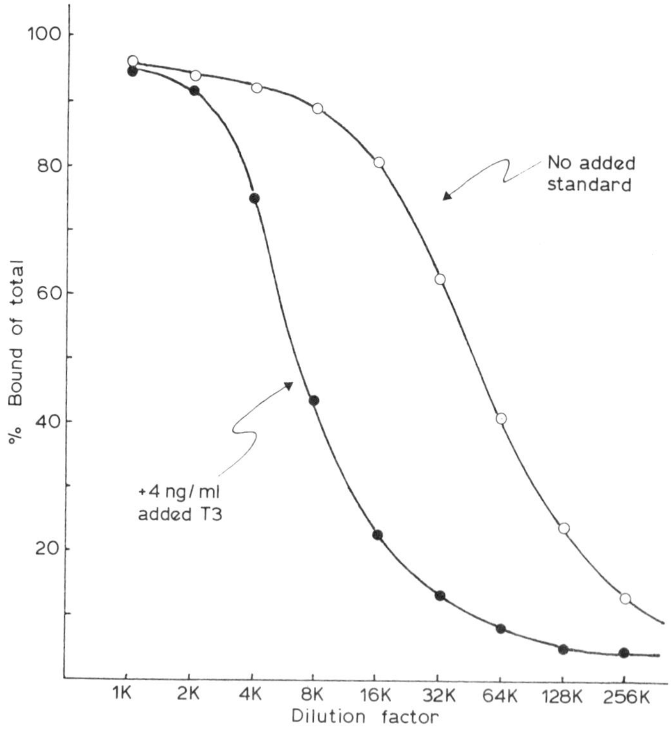

Fig. 4. Antiserum titration curves for an anti-triiodothyronine antiserum, batch no. HP/S/60-IIA: ○-○, antiserum dilution curve; ●-●, antiserum displacement curve.

the most concentrated solutions of the antiserum and the lowest binding at the highest dilution. A line is then drawn parallel to the abscissa from the point which represents 50% of the *maximum bound* label (not 50% of the total label added to the system). The point where this line intersects the dilution curve is the titre of the antiserum.

Avidity

It is possible to obtain an indirect assessment of avidity of an antiserum at the same time as one is setting up the antiserum dilution curve to determine its titre. This additional information is provided from an antiserum *displacement* curve. This curve is obtained by incubating a set of tubes whose composition is identical to those used to derive the antiserum dilution curve, except that an aliquot of the unlabelled antigen is added to each one. This can either be the top standard to be used in the standard curve, or if one is particularly concerned with selecting very sensitive antisera, one can choose a very low concentration of unlabelled antigen, similar to the expected detection limit of the assay. This second set of tubes will allow a similar sigmoid curve to be constructed (Fig. 4), whose linear portion is parallel to that of the antiserum dilution curve, but displaced to the left of it. The degree of displacement horizontally is an indirect measure of the avidity of the antiserum; the greater the displacement, the greater the avidity. So in one operation, one can obtain an estimate of the titre and an indirect assessment of avidity.

This approach can also be extended to include an assessment at this stage of the specificity of an antiserum in respect of its cross-reaction with the most likely interfering compound. At the same time that the antiserum dilution and antiserum displacement curves are set up, a third curve can be included, the composition of the incubates being similar to those of the antiserum displacement curve, except that instead of adding an aliquot of the unlabelled specific antigen, one includes a similar volume of the likely cross-reactant. These tubes are incubated and treated in a similar manner to those of the other two curves and the percentage binding of the labelled antigen calculated as before. These values are then plotted on the same graph as the antiserum dilution and antiserum displacement curves. The closer this third curve is to the antiserum *dilution* curve, the more specific the antibody under examination. If the antibody exhibits no cross-reaction, the cross-reactivity curve will be superimposable with the antiserum dilution curve.

It should, however, be borne in mind that this indirect method of comparing the avidity of different antisera does not allow one to determine its numerical value in terms of litres/mole. Avidity is defined as the energy with which the combining sites of an antibody bind its specific antigen. It is essentially the same as the association constant K_A in physical chemistry; with:

$$K_A = \frac{[Ag Ab]}{[Ag][Ab]}$$

where [AgAb], [Ag] and [Ab] are the molar concentrations of the antigen–antibody complex, free antigen and free antibody, respectively.

If it is necessary to know the avidity of an antiserum, it can be obtained in a number of ways, the principal one being by means of a Scatchard plot (Scatchard, 1949). This involves plotting the ratio of the bound to free labelled antigen on the ordinate against the total molar concentration of bound antigen, i.e. labelled + unlabelled antigen, on the abscissa. The slope of such a plot is $-K$, which has the dimensions of litres/mole. The intercept on the abscissa gives n, the total number of binding sites present, and that on the ordinate gives the product Kn. However, Walker (1977) has shown that there are a number of points to bear in mind when applying data to Scatchard analysis. All concentration units must be expressed in terms of molar concentration in the final equilibrium mixture *before* the separation agent is added. The proportion of the labelled antigen present in the bound fraction must be included, and to do this one must know the *weight* of labelled antigen added to the system. The labelled and unlabelled antigens must be allowed to react with antibody for the same length of time and the reaction must proceed to equilibrium. However, Klotz (1982) has recently drawn attention to the danger of using the Scatchard plot to determine the number of binding sites present, although it is a reliable method for measuring the binding constant, provided that there is only one binding site per molecule.

Specificity

In the third stage of antiserum assessment, it is necessary to determine whether or not an antibody is specific for its antigen. If an antibody is completely specific, it will only combine with that antigen, and not with any related compounds. In actual practice, this is rarely the case, since most antibodies do cross-react to a greater or lesser extent with analogues, metabolites, fragments and molecules possessing a similar amino acid sequence. The specificity of an antiserum should, therefore, be verified in three different ways: (a) by cross-reactivity curves, (b) by parallelism studies, and (c) the absence of noise factors.

The first of these procedures is concerned with determining the extent to which an antiserum cross-reacts with similar compounds by constructing cross-reactivity curves (Fig. 5). First one derives a standard curve for

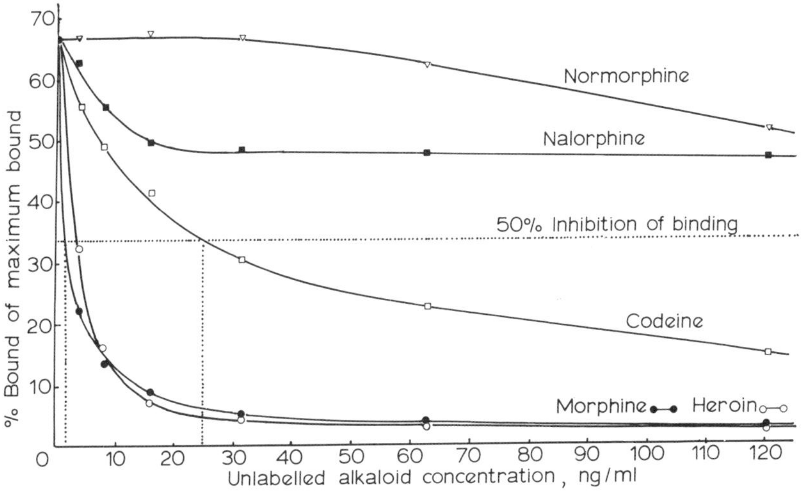

Fig. 5. Cross-reactivity curves for an anti-morphine antiserum, G/R/18-VIIA, immunogen: normorphine-BSA, label: 1 pM ^{3}H-dihydromorphine/tube.

the specific antigen by incubating a constant, but limited, amount of antibody with a constant amount of labelled antigen in the presence of increasing concentrations of the specific antigen. At the same time and in the same assay, one incubates with the same quantity of antibody and labelled antigen, increasing concentrations of possible cross-reacting compounds. After a period of incubation, the amount of labelled antigen in each tube is determined. The results are then plotted in a series of standard curves. Figure 5 shows the cross-reaction of a morphine antiserum with its specific antigen, as well as with other similar alkaloids. One can see that heroin cross-reacts with the antiserum in a similar manner to morphine. Some of the other alkaloids cross-react less completely and one is then faced with the task of expressing the extent to which they do so quantitatively. Cross-reaction is arbitrarily defined as the ratio of the weight of the specific antigen required to reduce binding of the label at zero by 50%, to the weight of the cross-reactant required to reduce binding of the label at zero by the same amount, multiplied by 100. The cross-reactivity of the morphine antiserum with codeine in the figure shown is just under 5%. It is thus possible to obtain an expression of the degree of interaction between the antiserum and other possible interfering materials.

One thing, however, should be stressed, and that is that the percentage cross-reaction of an antiserum with a particular compound is not a constant value, and it cannot be used as a correction factor. Because this value is determined in the complete absence of unlabelled specific antigen, it represents a maximum value, which is likely to be reduced as the concentration of specific antigen in the sample is increased from zero. It has been found with some steroid antisera that the percentage cross-reactivity of a particular analogue may fall by as much as half as the level of the specific antigen in the incubation mixture is raised to physiological levels. One way of overcoming this problem is to determine the degree of cross-reaction in the presence of low, medium and high levels of the specific antigen.

Another objection to the conventional determination of cross-reactivity is that the level of cross-reaction also depends to a certain extent on the level of binding in the zero standard tube. Ekins has suggested that cross-reaction should be expressed in terms of potency estimates (Ekins *et al.*, 1968) which are derived from the ratio of the avidities of the cross-reactant and the specific antigen.

One final point about cross-reactivity values is that they are affected considerably by the length of time for which the reaction mixture has been allowed to incubate. Vining *et al.* (1981) showed that a number of steroid hormone antisera become more specific the longer the incubations are continued. The cross-reaction of a testosterone antiserum with its principal metabolite, dihydrotestosterone, was progressively reduced from 42·3% after 10 min incubation to 11·3% after 16 h. This observation explains why laboratories, using the same reagents or kit, but different incubation times, obtain different values for the analyte concentration of the same quality assurance sample.

From the foregoing, it must be apparent that a particular cross-reactivity value only relates to the system in which it was determined. Therefore, it should not be assumed that the cross-reactivity of an antibody in an ELISA system will be the same as it is in a radioimmunoassay.

This method of checking the specificity of an antiserum, by determining the percentage cross-reaction from cross-reactivity curves, has one important limitation. It can only be applied to those compounds which are known to exist and for which a pure preparation is available. Unfortunately, the precursor, metabolite and fragment forms of a specific antigen are not always known and so any possible cross-reaction between them and the antiserum being assessed cannot be checked by this method. There is, however, another procedure available which enables this to be achieved as well as at the same time checking for the presence of other components in the sample which may cross-react. It involves comparing a standard curve of the specific antigen with a curve obtained by double diluting, with the same matrix as is used to dilute the standards, a sample containing a high level of the specific antigen. If the two curves are superimposable at all points, then one can assume that the antiserum being examined is specific for the antigen in question and is not susceptible to interference from other related compounds which may be present in the sample. This is the test of parallelism, which, since the dose response curves are curvilinear, is sometimes referred to as generalised parallelism.

The final check on the specificity of an antiserum is to eliminate the possibility of it giving false positive values when there is no specific antigen present. In immunoassay, this phenomenon is usually caused by 'noise factors' in the antiserum. It is possible to check for their presence by assaying samples known to contain no, or negligible, amounts of the specific antigen. Any spurious false positive result is likely to be due to the antibody recognising and cross-reacting with a component of the matrix other than the specific antigen.

ASSAY CONDITIONS

As mentioned at the beginning of this chapter, the extent of the interaction between antigen and antibody in immunoassay is considerably influenced by the environment in which the reaction occurs. For this reason, the assay conditions should be carefully optimised. Since one is dealing with such small quantities of analyte in immunoassay, phenomena which do not normally cause any problems when measuring milligram and microgram quantities of analyte, become increasingly important when levels of only nanogram and picogram amounts are encountered. The principal problem at these latter levels of analyte concentration is adsorption of material to the walls of the assay tube and the tips of the dispensing apparatus. It is for this reason that a protein is usually added to the assay buffer to block the binding sites responsible. The most frequently used one is bovine serum albumin (BSA) at a concentration usually between 0·1 and 0·5%. Care should be taken, in the choice of this material, to avoid preparations which contain proteolytic enzymes, since these may cause degradation of the antigen being measured. Some preparations of BSA are now marketed specifically for immunoassay use. BSA is used principally in assays for protein and polypeptide antigens. Like all albumins however, it has a high affinity for small molecules such as steroids and drugs. In those assays where albumin is known to interfere, the relatively inert protein gelatin should be used instead, after first checking that it is a suitable alternative.

The phosphate buffer used in most assays is an excellent nutrient medium for a number of micro-organisms. It is therefore usual to add a bacteriostat, such as 0·2% merthiolate or 0·1% sodium azide to the buffer, to prevent their growth, even if the assay incubation time is only of short duration.

One of the principal lessons which the early immunoassayists learnt to their chagrin was that one cannot usually dilute standards in buffer when the analyte, whose concentration it is wished to measure, is present in a different matrix, because of the effect the matrix may have on the antigen–antibody reaction. The matrix is defined as those substances, other than the analyte, which are present in the sample or sample extract. The standards in an immunoassay should, therefore, be diluted in antigen-free matrix, unless it has been demonstrated beyond doubt that the matrix of the samples being analysed has no effect on the antigen–antibody reaction in question.

In clinical medicine, the need for antigen-free matrices poses some difficulties, since many of the substances being measured, e.g. hormones, are of endogenous origin. Several different methods have been developed for preparing these materials. They all involve 'stripping' the analyte in question from a bulk quantity of the matrix being assayed, e.g. plasma or serum. Some of them, such as the immunoadsorbents and, to a lesser extent, the lectins, are specific for the analyte being removed, whilst others, such as charcoal, which is used for small molecular weight substances, remove many compounds in addition to the analyte.

However, in food analysis, one is frequently checking for the presence of contaminants or residues which are not normally present in the materials being assayed. In this case, it is possible to use material from, for example, uncontaminated plants or untreated animals (Chapter 13), as the antigen-free matrix.

PHASE SEPARATION

In most assays (the heterogeneous ones) at the end of the incubation period, it is necessary to separate the labelled antigen which has bound to antibody (the bound fraction) from that which has not (the free fraction). Since the reaction between the antigen and its specific antibody does not produce a precipitate, it is necessary to effect a physical separation of the two forms before one or the other can be measured. This process is termed phased separation. The ideal separation system should allow complete separation of one fraction from the other, yet, at the same time, should not disturb the equilibrium between them. This proviso is most important since there is no point in allowing a system to incubate for a long period of time in order to obtain the maximum binding, if the phase separation system one uses is going to disturb that equilibrium. Such a disturbance is usually seen with the adsorptive group of phase separation reagents, particularly when the antibody being used is of low avidity.

Most phase separation methods work by creating a precipitate of one of the fractions, which can then be separated from the other by centrifugation. When one produces the precipitate of one form, either the bound or free, there is every possibility that some of the supernatant phase will be trapped within the precipitate. This is termed 'misclassification' and is one of the reasons why sometimes it is difficult to obtain really low detection limits. The effect of this contamination is to increase the level of non-specific binding in an assay. However, high levels of non-specific binding are not necessarily due to misclassification. The small quantity of supernatant trapped within the precipitate can sometimes be removed by washing after the supernatant has been aspirated. In order to do this without any loss, the precipitate needs to be reasonably particulate.

There are a number of methods available for effecting phase separation and these are shown in Table 2. These can be divided into five basic groups: those embodying (a) chromatography; (b) chemical precipitation; (c) adsorption; (d) immunology; (e) solid phase. All but the first group have achieved widespead use, since there is no ideal phase separation method which is suitable for all antigens, although the second antibody technique now vies with the solid phase systems for the title.

Chromatoelectrophoresis is included in the list for historical reasons only, since it is the method which Berson and Yalow used in the first radioimmunoassay. Samples of the incubates are applied near the ends of separate strips of chromatography paper and these are subjected to a

Table 2
Phase Separation Methods

Chromatography	Chromatoelectrophoresis Gel filtration Wick (paper chromatography)
Chemical precipitation	Na_2SO_4, $(NH_4)_2SO_4$ Ethanol Polyethylene glycol
Adsorption	Coated charcoal Silica — QUSO G32, Vycor, Spherosil Talc
Immunology	Double or second antibody Protein A
Solid phase	Coated tube Coated disc Immunosorbent Magnetic particles

combination of ascending chromatography and electrophoresis. The antibody-bound fraction migrates whilst the free fraction remains at the origin. Each strip is then scanned in a radiochromatogram scanner and the proportion of the label which was bound to antibody is determined.

It is no small wonder, therefore, that alternative methods to this very tedious technique were sought and developed. Chemical precipitation with either ammonium sulphate or sodium sulphate is a classical method used by immunologists to prepare the γ-globulin fraction of an antiserum. It was found that these two chemicals could be used to precipitate the antibody-bound fraction of an incubate without disturbing the equilibrium. The separation is reasonably rapid and the reagents relatively cheap. However, the method does suffer from two serious disadvantages. The precipitate produced is composed of aggregated material which tends to trap some of the free phase, giving rise to serious misclassification. This can be reduced by washing the precipitate with a 40% saturated solution of ammonium sulphate before counting. Alternatively, an aliquot of the supernatant can be removed and counted instead, as is done with assays using tritium-labelled antigens. The other disadvantage is that ammonium sulphate cannot be used in the assays for large protein molecules, on account of the co-precipitation it causes of the labelled antigen. Neither can it be used with enzyme-labelled antigens.

Turning to the adsorption methods, by far the most important is coated charcoal, which was introduced by Herbert *et al.* (1965) for use in radioimmunoassay. In their original method, an activated charcoal, Norit A, was coated with dextran which, in turn, acted as a molecular sieve. Dextran is permeable to small molecules, but excludes large ones. When dextran-coated charcoal is added to the assay tube at the end of the incubation period, the activated charcoal core adsorbs the free labelled antigen, providing it is small enough to penetrate the dextran coating. The antibody-bound antigen will be excluded on account of its much larger molecular size and therefore remain in solution. The final separation is achieved by centrifugation, which permits a complete separation of the bound and free phases.

It is now known that this is a very elementary explanation of the principle involved, since a number of other coatings can be substituted for dextran which do not possess molecular sieving properties. Albumin, gelatin, and even serum, have been used in its place. The method is very simple to use and the separation is immediate, although in actual practice it is usual to leave the charcoal in contact with the incubate for about 10 min before centrifuging the assay tubes.

The disadvantage of the method is that it is only suitable for small molecules, such as steroids, drugs and peptides up to a molecular weight of approximately 10 000 daltons. This is because above this molecular weight range, the coating begins to exclude both the free and the bound forms of the labelled antigen. Another serious disadvantage is that the charcoal may disturb the equilibrium attained between the free and bound fractions at the end of the incubation period, by stripping off the antigen bound to antibody. This is particularly true of low avid antibodies which are usually produced in the period following the priming injection. There have been several instances where the appearance of antibodies in a newly immunised animal has been completely missed because coated charcoal was used as the phase separation medium. It can be argued that these antibodies are not worth having, so nothing has been lost. On the other hand, it is sometimes important to know whether an animal has responded at all, and in this situation it would be better to use another phase separation method, such as ammonium sulphate or second antibody.

Because of its ability to strip antigen from the bound complex, two precautions should be taken when using coated charcoal. The first is that the separation should be carried out at 4°C in ice-water, so as to reduce the rate of dissociation of the antigen–antibody complex, and hence the deleterious effect that charcoal may have on the final equilibrium. The second is that the coated charcoal should be left in contact with the incubate for the minimum time possible, for the same reason.

The second or double antibody method, which was first developed by Utiger *et al.* (1962) for effecting phase separation of protein hormone radioimmunoassays, is rapidly becoming the method of choice for an even wider range of assays, from those for small molecules, such as haptens, to enzyme-labelled immunoassays. The aim of this separation method is to create a sufficiently large immunological complex, or micelle, incorporating the first antibody-specific antigen complex, so that it is possible to separate the bound phase by ordinary centrifugation.

In an attempt to increase the overall weight and density of the complex, Utiger and his co-workers conceived the idea of making the first antibody become the antigen of a second antigen–antibody reaction. To achieve this, it was necessary to raise an antibody to the first antibody in another species to that in which the first antibody was raised. If the first antibody was raised in a rabbit, then γ-globulins from normal rabbits are injected into an animal of another species, such as a donkey, to produce a donkey anti-rabbit γ-globulin antiserum. Such an antiserum will react with the Fc fragment of any rabbit γ-globulins, whether or not specific for the first antigen.

This second antibody is normally added at the end of the incubation period of the first antibody with its specific antigen, and the reaction mixture allowed to incubate for a further period of between 8 and 24 h. At the end of this time, all that is necessary is centrifugation of the assay tubes to compact the precipitated first–second antibody complex at the bottom of the tube. The supernatant is aspirated before the amount of labelled antigen is determined in pelletted bound fraction.

Although this is a simple procedure, there are a number of points which should be borne in mind if this method is to be used successfully. Not all anti-γ-globulin antisera are suitable for use as precipitating second antibodies in immunoassay, and so it is essential to check the ability of each batch of reagent to do this. The amount of precipitate actually produced by the reaction of the second antibody with the first antibody to the specific antigen being measured, is normally very small, so small that it can hardly be seen. This causes problems in compacting the material into a pellet, which in turn gives rise to loss of precipitate at the separation stage and hence loss of precision in the assay. However, this difficulty can be overcome by the addition to the assay tubes of 'carrier serum', which is normal serum from the same species as that in which the first antibody was raised. This contributes extra γ-globulin to the system, which, in turn, increases the bulk of the final precipitate, and results in more complete precipitation of the specific antigen–antibody complex. However, when the first antibody is used at a final concentration of less than 1:3000, the non-specific γ-globulins in the first antiserum are usually present in sufficient concentration to provide the extra bulk necessary, and the addition of carrier serum is, in this case, not required.

It goes without saying that the use of carrier serum increases the amount of second antibody required to ensure complete precipitation of the first antibody–antigen complex. It is therefore necessary to optimise each assay system with regard to the amount of second antibody and carrier serum that must be added. This requires that the titration of second antibody must be performed simultaneously with the titration of carrier serum and the process repeated each time a new batch of reagent is introduced into the assay system. The formation of the first antibody–second antibody precipitate is additionally susceptible to interference from complement. This can be inactivated by the incorporation of the disodium salt of ethylene diamine tetra-acetic acid (EDTA) into the buffer used to dilute the second antibody and carrier serum. It should not be present during the first antibody–antigen reaction, since it is known to inhibit this process.

Despite these difficulties, the second antibody phase separation procedure is of universal applicability since it is not limited by the molecular weight of the specific antigen. Because the second antibody primarily reacts with the Fc fragment of the first antibody, it does not disturb the equilibrium between the first antibody and its specific antigen, which means that it can be used with low avid antibodies. Until recently, its more general application has been limited by the lengthy time interval required for the precipitate to be formed, thus making it unsuitable for use in assays for small molecules, which come to equilibrium relatively quickly.

This problem has now been overcome by the use of accelerators of the precipitate formation. The interaction of the second antibody with the first antibody–antigen complex to produce a precipitate takes place in two quite separate stages. The first of these is the antigen–antibody reaction between the first and second antibodies. This is quite fast, taking between 20 to 30 min to complete. It does not, however, result in the formation of a precipitate, which is the second stage of the process. To bring this about, it is necessary for several first antibody–second antibody complexes to aggregate and form an insoluble micelle by loss of their water of hydration. It is this stage which takes the time.

Martin & Landon (1975) showed that it was possible to accelerate this loss of water of hydration, and hence increase the rate of precipitate formation, by including ammonium sulphate, at a final concentration in the assay tube of 1·2 M, with the second antibody. The whole separation period was then reduced to 30 min. The accelerator not only decreases assay time, but also allows the second antibody to be used more conservatively in a more dilute form.

We ourselves prefer to add an equal volume of 4% polyethylene glycol 6000 (PEG 6000) to the reaction mixture after it has incubated for 30 min with the second antibody and then to centrifuge the assay tubes 10 min later. This approach is derived from an observation by Hellsing (1972) that the addition of 4% PEG 6000 in immunonephelometry, not only considerably reduced the reaction time, but also increased the analytical range and improved the sensitivity, the latter by a factor of five- to eight-fold for many proteins. Since then, Edwards (1983) has undertaken a critical evaluation of this accelerator in radioimmunoassay and found that it significantly reduced the level of non-specific binding, compared to those assays using the conventional non-accelerated second antibody method.

The minuteness and solubility of the specific antigen–antibody complexes in immunoassay have been the source of many problems, particularly at the phase separation stage. In order to overcome these difficulties, as well as to provide a means of automating the technique, the solid-phase group of separation methods was developed.

The first of these was the coated tube method of Catt & Tregear (1967). In this approach, the specific antibody is adsorbed non-covalently to the walls of polystyrene assay tubes by incubating a pre-determined, but limiting, amount of antiserum in them overnight. The antibodies bind non-specifically to the walls of the tube and are retained, immobilised on its surface, when the contents are emptied and the tube washed. The assay is performed by adding labelled antigen and unlabelled antigen to the coated tube and incubating in the normal way. Phase separation is achieved merely by either emptying the incubate from the tube and washing the latter before counting the bound phase, or removing an aliquot of the incubate to determine the amount of free phase. It should be stressed that this is still a limited reagent assay, although the same approach is used in the ELISA technique to coat the wells of microtitre plates.

The advantages of the method are obvious: centrifugation of the assay tubes is eliminated, and separation is immediate. Large numbers of tubes can be coated in one operation and stored until required. However, the method, although admirably suited to screening procedures, lacks adequate precision for quantitative work, principally because of the variability in the amount of antibody immobilised during the coating stage. Not only do tubes from different batches of the same manufacturer vary, but individual tubes of the same batch do so as well. Added to this is the additional problem that the immobilised antibodies tend to 'leak' off during storage and during the assay.

Obviously one way to overcome this last difficulty is to attach the antibodies covalently to a solid phase and this was the approach used by Wide *et al.* (1967) who coupled the γ-globulin fraction of the specific antiserum to small dextran particles (Ultrafine Sephadex), using cyanogen bromide. The immobilised antibodies are stable on storage for quite long periods of time. The technique is not so convenient as the coated tube method to perform, since the antibody reagent needs to be kept agitated during dispensing, to ensure that the aliquots are uniform in quantity of antibody, and also during the incubation period. In addition, the tubes need centrifuging before the phases can be separated.

However, this approach has been the basis of the methods developed by most commercial kit manufacturers, who have used a wide range of materials for the solid phase. These have ranged from small beads of glass, polystyrene, microcrystalline cellulose and latex to, more recently, polymer-coated magnetic iron particles (Nye *et al.*, 1976). They should be referred to more correctly as magnetisable particles, since they respond to an applied magnetic field, but are not themselves intrinsically magnetic. They are extremely versatile, since they can also be used with second antibodies and adsorbents, such as charcoal, to provide universal phase separation reagents, although this property is not exclusive to magnetic particles themselves. Phase separation with these particles merely involves placing the assay tubes, after incubation, in a holder, over a strong permanent magnet and decanting the supernatant. The immobilised antibody-bound fraction is retained at the bottom of the tube.

The use of solid-phased second antibodies has several advantages over their first antibody counterparts, irrespective of the nature of the support medium being used. Not only do they provide a universal reagent for use in several different assays using a first antibody raised in the same species, but, because it is an excess reagent, they do not require precise dispensing like solid-phased first antibodies, whose concentration in the assay system is critical. A further advantage of solid-phased second antibody is that it does not slow the rate of association of the first antibody–antigen complex, as sometimes happens when the first antibody is attached to solid phase.

The preparation of solid-phased antibodies for use in immunoassay shares many common features with the preparation of these reagents for immunometric assay, the labelled antibody variant.

IMMUNOMETRIC ASSAY

What has been discussed so far is classical immunoassay, a system which uses a labelled form of the antigen being measured and a limited amount of specific antibody. It is difficult to prepare labels of some antigens, because they are not stable enough to withstand either the conditions of radioiodination or those forming a covalent bond with a non-isotopic label. It was primarily to overcome these problems that Miles & Hales (1968) conceived the idea of labelling the antibody instead of the antigen, and using this reagent in what they called an immunometric assay.

This type of assay differs from classical immunoassay in that the analyte is assayed directly by combination with the labelled antibodies, rather than in competition with a labelled derivative for a limited amount of antibody. To perform the assay, the antigen being measured is incubated with an excess of the labelled specific antibody, as is shown in Fig. 6. At the end of incubation, the excess antibody remaining is separated from the antigen–antibody complex by adding antigen which has been coupled to a solid phase. After a further short period of incubation, the tubes are centrifuged and the activity of the labelled antibody in either the precipitate or the supernatant determined. Measuring the activity in the liquid phase has the advantage that it is directly related to the original antigen concentration in the sample or standard solution.

The assay therefore consists of three separate stages: (a) isolation of the specific antibodies, (b) labelling the isolated antibodies, and (c) the assay itself, which has already been described. In order to develop assays with comparable or superior sensitivity than classical immunoassay, it is important to label only the antibodies which are specific for the antigen

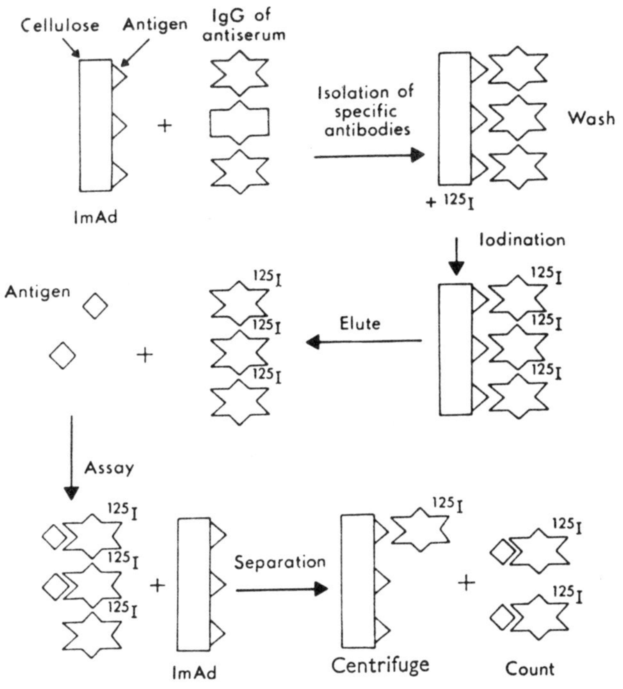

Abbreviation:
ImAd: immunoadsorbent

Fig. 6. Diagram of the stages involved in the immunoradiometric assay. (Reproduced with permission from Woodhead *et al.*, 1974.)

being measured, and not the whole γ-globulin fraction. This requires that the specific antibodies are first isolated from the neat antiserum with an immunoadsorbent of the antigen coupled to solid phase. The immunoadsorbent is left in the antiserum for several days in order to obtain the maximum number of high avidity antibodies. After removing the immunoadsorbent by centrifugation, it is washed well in order to remove any serum proteins present other than the specific antibody, which is then ready for labelling. Any of the labels that have been used in classical immunoassay can also be used in immunometric assay, although possibly the two most commonly used ones to date are ^{125}I and enzymes. The labelling procedure is performed with the antibody fraction still immobilised on the immunoadsorbent. This has the advantage of protecting the antigenic binding site on the antibody from being damaged or blocked by the label. The immunoadsorbent is then washed to remove the labelling reagents and products.

The labelled antibodies are eluted from the solid phase by first washing with a pH 3 buffer to elute the low avid antibodies and then with a pH 2 buffer to elute the higher ones. The labelled preparation is best stored, however, deep frozen on the immunoadsorbent, and the labelled antibodies eluted just prior to use. The immunoglobulins are very robust molecules and it is not surprising, therefore, that iodinated antibodies have a shelf life of between two and three months.

A variation of the immunometric assay is the two-site or sandwich assay, described by Addison & Hales (1971), which was developed for measuring antigens which had at least two quite separate antigenic determinants. In this procedure, the specific antibody to the first antigenic determinant is immobilised on a solid phase, as is shown in Fig. 7.

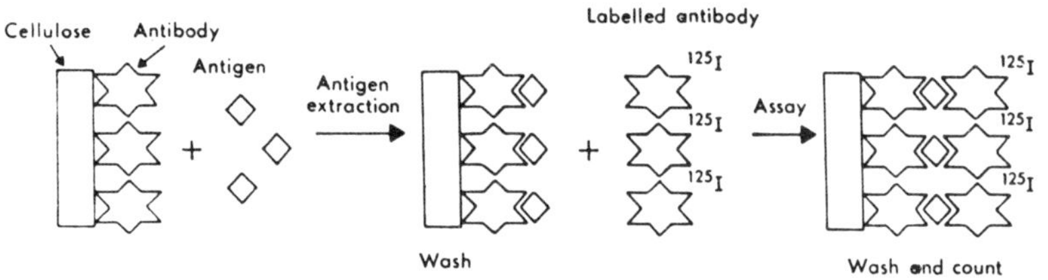

Fig. 7. Diagram of the two-site assay procedure. (Reproduced with permission from Woodhead *et al.*, 1974.)

This is then incubated in excess with the sample or standard containing the antigen, which combines with the immobilised antibody. The immunoadsorbent is then precipitated by centrifugation and washed before adding an excess of the labelled specific antibody to the second antigenic determinant on the molecule. After a further period of incubation, the immunoadsorbent is again precipitated by centrifugation with the antigen sandwiched between the antibody on the solid phase to the first determinant and the labelled antibody to the second determinant. The activity of the label in the precipitate is then determined and is directly proportional to the antigen content of the sample.

It will be apparent from the foregoing that the labelled antibody will only be associated with the immunosorbent complex if the antigen which is trapped between it and the antibody on the solid phase carries both antigenic determinants. If the molecule only possesses the second antigenic determinant, it will not bind to the immunosorbent, and if it only carries the first determinant, the labelled antibody will not become associated with the immunosorbent complex. The two-site method, therefore, offers much greater specificity than classical immunoassay because it recognises two separate sites on the molecule simultaneously. It thus has the potential for discriminating against precursors, prohormones, metabolites and fragments, if these only possess one of the determinants and not the other. However, a prerequisite for this particular assay is the existence of two antisera containing antibodies directed against the two different determinants on the molecule. Although it is possible to produce polyclonal antibodies, which are capable of achieving this, it has become a much more practical proposition with the advent of monoclonal antibodies. One could almost say that monoclonal antibodies were created for two-site assays.

One further modification of the two-site assay is the system described by Beck & Hales (1975), which used a labelled second antibody, instead of a labelled first antibody. The second antibody is added after the antigen has been sandwiched between the two first antibodies. For this system to succeed, however, it is necessary to raise the two first antibodies to the antigen being measured in phylogenetically distinct species; otherwise the labelled second antibody will also interact with uncomplexed antibodies on the solid phase when no antigen is present.

The similarity between these two-site assay systems and the noncompetitive ELISAs will be seen in Chapter 3.

REFERENCES

ADDISON, G. M. & HALES, C. N. (1971) Two-site assay of human growth hormone. *Hormone and Metabolic Research*, **3**, 59–60.

BECK, P. & HALES, C. N. (1975) Immunoassay of serum polypeptide hormones by using ^{125}I-labelled anti-(immunoglobulin G) antibodies. *Biochemical Journal*, **145**, 607–16.

CATT, K. & TREGEAR, G. W. (1967) Solid phase radioimmunoassay in antibody-coated tubes. *Science*, **158,** 1570–2.

EDWARDS, R. (1983) The development and use of PEG assisted second antibody precipitation as a separation technique in radioimmunoassays. In: *Immunoassays for Clinical Chemistry*, Hunter, W. M. & Corrie, J. E. T. (eds), Churchill Livingstone, Edinburgh, pp. 139–46.

EKINS, R. P., NEWMAN, G. B. & O'RIORDAN, J. L. H. (1968) Theoretical aspects of 'saturation' and radioimmunoassay. In: *Radioisotopes in Medicine: in vitro*

Studies, Hayes, R. L., Goswitz, F. A. & Murphy, B. E. P. (eds), US Atomic Energy Commission, Oak Ridge, Tennessee, pp. 59–100.

FRAKER, P. J. & SPECK, J. C. (1978) Protein and cell membrane iodinations with a sparingly soluble chloramide, 1,3,4,6-tetrachloro-3α,6α-diphenylglycoluril. *Biochemical and Biophysical Research Communications*, **80**, 849–57.

GILBY, E. D. & JEFFCOATE, S. L. (1973) Studies on the affinity of a testosterone antiserum for testosterone and various conjugated derivatives. *Journal of Endocrinology*, **57**, x/vii.

GREENWOOD, F. C., HUNTER, W. M. & GLOVER, J. S. (1963) The preparation of ^{131}I-labelled human growth hormone of high specific radioactivity. *Biochemical Journal*, **89**, 114–23.

HELLSING, K. (1972) In: *Colloquium on ATP*, Technicon Instruments Corp., Tarrytown, NY, p. 17, quoted by Ritchie, R. F. (1975).

HERBERT, V., LAU, K. S., GOTTLIEB, C. W. & BLEICHER, S. J. (1965) Coated charcoal immunoassay of insulin. *Journal of Clinical Endocrinology*, **25**, 1375–84.

HURN, B. A. L. (1971) Discussion on Antisera, In: *Radioimmunoassay Methods*, Kirkham, K. E. & Hunter, W. M. (eds), Churchill Livingstone, Edinburgh, pp. 181–3.

JEFFCOATE, S. L. (1981) *Efficiency and Effectiveness in the Endocrine Laboratory*, Academic Press, London, 223 pp.

KLOTZ, I. M. (1982) Numbers of receptor sites from Scatchard graphs: Facts and fantasies. *Science*, **217**, 1247–9.

MARCHALONIS, J. J. (1969) An enzymic method for the trace iodination of immunoglobulins and other proteins. *Biochemical Journal*, **113**, 299–305.

MARTIN, M. J. & LANDON, J. (1975). The use of ammonium sulphate and dextran together with second antibody at the separation stage in radioimmunoassay. In: *Radioimmunoassay in Clinical Biochemistry*, Pasternak, C. A. (ed.), Heyden, London, pp. 269–81.

MILES, L. E. M. & HALES, C. N. (1968) Labelled antibodies and immunological assay systems. *Nature*, **219**, 186–9.

NYE, L., FORREST, G. C., GREENWOOD, H., GARDNER, J. S., JAY, R., ROBERTS, J. R. & LANDON, J. (1976). Solid phase magnetic particle radioimmunoassay. *Clinica Chimica Acta*, **69**, 387–96.

RITCHIE, R. F. (1975) Specific protein assay. *Federation Proceedings*, **34**, 2139–44.

ROBINSON, J. D., AHERNE, G. W., TEALE, J. D., MORRIS, B. A. & MARKS, V. (1975*a*). Problems encountered in the use of radioiodination tags in radioimmunoassays for drugs. In: *Radioimmunoassay in Clinical Biochemistry*, Pasternak, C. A. (ed.), Heyden, London, pp. 113–23.

ROBINSON, J. D., MORRIS, B. A., PIALL, E. M., AHERNE, G. W. & MARKS, V. (1975*b*) The use of rats in the screening of drug-protein conjugates for immunoreactivity. In *Radioimmunoassay in Clinical Biochemistry*, Pasternak, C. A. (ed.), Heyden, London, pp. 101–11.

SALACINSKI, P., HOPE, J., MCLEAN, C., CLEMENT-JONES, V., SYKES, J., PRICE, J. & LOWRY, P. J. (1979) A new simple method which allows theoretical incorporation of radio-iodine into proteins and peptides without damage. *Journal of Endocrinology*, **81**, 131P.

SCATCHARD, G. (1949) The attractions of proteins for small molecules and ions. *Annals New York Academy of Sciences*, **51**, 660–72.
THORELL, J. I. & JOHANSSON, B. G. (1971) Enzymatic iodination of polypeptides with ^{125}I to high specific activity. *Biochimica et Biophysica Acta*, **251**, 363–9.
UTIGER, R. D., PARKER, M. L. & DAUGHADAY, W. H. (1962) Studies on human growth hormone. I. A radioimmunoassay for human growth hormone. *Journal of Clinical Investigation*, **41**, 254–61.
VINING, R. F., COMPTON, P. & MCGINLEY, R. (1981) Steroid radioimmunoassay — effect of shortened incubation time on specificity. *Clinical Chemistry*, **27**, 910–13.
WALKER, W. H. C. (1977) An approach to immunoassay. *Clinical Chemistry*, **23**, 384–402.
WIDE, L., AXEN, R. & PORATH, J. (1967) Radioimmunosorbent assay for proteins. Chemical couplings of antibodies to insoluble dextran. *Immunochemistry*, **4**, 381–6.
WOODHEAD, J. S., ADDISON, G. M. & HALES, C. N. (1974) The immunoradiometric assay and related techniques. *British Medical Bulletin*, **30**, 44–9.
YALOW, R. S. & BERSON, S. A. (1959) Assay of plasma insulin in human subjects by immunological methods. *Nature*, **184**, 1648.
YALOW, R. S. & BERSON, S. A. (1960) Immunoassay of endogenous plasma insulin in man. *Journal of Clinical Investigation*, **39**, 1157–75.

3

Principles of Enzyme Immunoassay

M. J. SAUER, J. A. FOULKES

Ministry of Agriculture, Fisheries and Food, Cattle Breeding Centre, Shinfield, Reading, UK

and

B. A. MORRIS

Division of Clinical Biochemistry, Department of Biochemistry, University of Surrey, Guildford, UK

INTRODUCTION

The development of radioimmunoassay (RIA) 25 years ago (Berson & Yalow, 1959) marked the beginning of a new era in biochemical analysis, offering the possibility of assaying a wide variety of proteins and polypeptides. Almost 10 years passed before the potential of this method for the determination of haptens was realised (Oliver *et al.*, 1968). With this came the very real possibility of assaying, with unprecedented sensitivity and specificity, any biochemical substance against which an antibody could be raised. Disadvantages, such as there are, revolve around the use of radioisotopes for quantification. Some isotopes have short half-lives, their determination requires sophisticated and expensive equipment and their toxic nature necessitates the application of strict regulatory controls. As a result, RIA has been excluded from many small laboratories, especially in the food manufacturing industry.

Although many alternative forms of labelling have been considered (Schall & Tenoso, 1981), the use of enzyme-labelled reagents for immunoassays (Engvall & Perlmann, 1971; van Weemen & Schuurs, 1971) has provided the most convenient means by which the sensitivity and specificity of RIA can be more generally applied. The reagents used are those commonly employed in general laboratories and are thus not associated with special hazards: the enzymes used for labelling are cheap and may be stored for in excess of one year at 4°C, or at room temperature when freeze-dried. Similarly, the equipment required for enzyme immunoassay (EIA) is common to the majority of the laboratories and requires only the availability of accurate pipettes and dispensers, and a colorimeter or spectrophotometer. End-point determination, i.e. the quantification of colour or fluorescence, need not be rate limiting. Microtitre plate readers enable automatic optical density (OD) determination at approximately one per second and microprocessor interfacing is possible for data reduction. Since EIA can be easy to perform and colorimetric end-points visually assessed, it has been adopted for simple screening procedures for use in the field (Voller *et al.*, 1976*a*).

ASSAY PRINCIPLES

Two basic types of immunoassay employ enzyme-labelled reagents: the enzyme immunoassay (EIA) and the sandwich assay, which are broadly analogous with RIA and immunoradiometric assay, respectively. They rely on the inherent characteristic of antigen and antibody to bind with high avidity and specificity.

Enzyme immunoassay

EIA encompasses immunoassays based on the 'saturation assay' principle in which the analyte is enzyme-labelled (Fig. 1). The assay comprises three components: (i) a limited and constant quantity of antiserum specific for the analyte, (ii) a limited and constant quantity of enzyme-labelled analyte, and (iii) standard quantities of analyte for calibration purposes (or unknown quantities of analyte in the test sample). When the

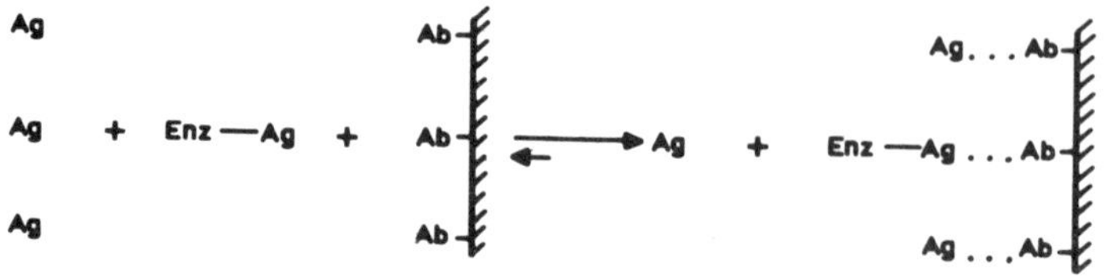

Fig. 1. Principle of enzyme immunoassay; Ag = antigen or hapten, Enz = enzyme label, ⨍ = solid phase, Ab = specific antibody.

three components of the system are mixed, labelled and unlabelled analyte compete for the limited number of antibody binding sites (Fig. 1). Since each test contains the same amount of antibody and labelled analyte, the greater the quantity of unlabelled analyte present, the smaller will be the quantity of labelled analyte binding to antibody. Separation of free labelled analyte from that bound to antibody, and subsequent addition of substrate to the bound fraction, allows enzyme activity to be measured and the degree of competition to be ascertained. The greater the quantity of analyte present, the fainter will be the colour produced. The use of standard amounts of analyte enables calibration curves to be constructed and quantification of analyte in the unknown sample by interpolation (Fig. 2). A typical EIA protocol is given in Table 1.

This type of EIA requires separation of free analyte from antibody-bound analyte for quantification and is termed 'heterogeneous'. The homogeneous EIA or enzyme–mediated immunoassay technique (EMIT) is a special case of EIA which does not depend on free and bound label separation to enable assessment of the degree of competition (Boguslaski & Li, 1982). Until recently, this method could only be applied to the determination of haptens. It relies on the use of enzyme–hapten conjugates in which enzyme activity is altered on binding with the antibody. Figure 3 illustrates the principle behind the most common form of EMIT in which enzyme activity is inhibited by specific interaction with the antibody: enzyme activity is proportional to the concentration of hapten since the degree of inhibition will be reduced by its presence. Although these assays can provide very rapid tests (2–3 min) since separation is not required, these procedures are far less sensitive than their heterogeneous conterparts (100–1000 fold) and their application has largely been limited to the monitoring of drugs during therapy.

Sandwich Assays

'Enzyme-linked immunosorbent assay' (ELISA) encompasses all solid-phase immunoassays using enzyme-labelled reagents, but is most commonly used to describe non-competitive solid-phase sandwich assays: it will, therefore, be used here in the more restricted latter context. These heterogeneous methods are analogous with immunoradiometric assays (Miles & Hales, 1968; Addison & Hales, 1971) and are based on the principle of 'reagent excess'. In contrast to EIA, there is no element of competition and the amount of enzyme label bound is directly proportional to the concentration of analyte. Solid-phase reagents are used

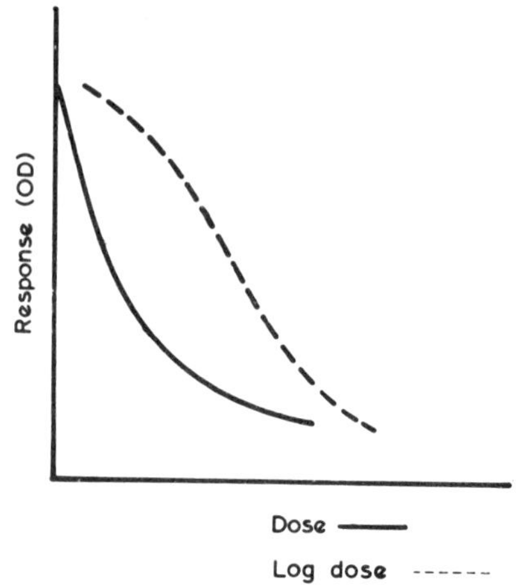

Fig. 2. Typical calibration curves for enzyme immunoassay.

Table 1
Typical Assay Protocol for Microtitre Plate Enzyme Immunoassay

1. Adsorb specific antibody onto wells of microtitre plate (Voller *et al.*, 1976*a*). This involves addition of diluted, purified antibody (200–300 μl, dependent on plate type) and incubation for 3 h at room temperature or overnight at 4°C. Store until required. Wells are emptied and washed prior to use.
2. Add sample or standard, then enzyme-labelled antigen (or hapten) to the wells and incubate for 2–3 h at room temperature.
3. Empty and wash wells with buffer (× 3).
4. Add substrate solution and incubate at a temperature suited to enzyme (20°C for peroxidase, 20–40°C for alkaline phosphatase or β-galactosidase). Stop reaction by addition of sodium hydroxide and determine absorbance. The reaction need not be stopped if an automatic microtitre plate reader is used since all 96 wells may be read in 1–$1\frac{1}{2}$ min.
5. Plot calibration curve and interpolate results.

for separation of free label from bound label which also facilitates the removal of excess reagents after each stage.

Although based on the same principle, there are numerous variants of the ELISA procedure (Schuurs & van Weemen, 1980), the majority of which use enzyme-labelled antibody as tracer: Table 2(a) illustrates the

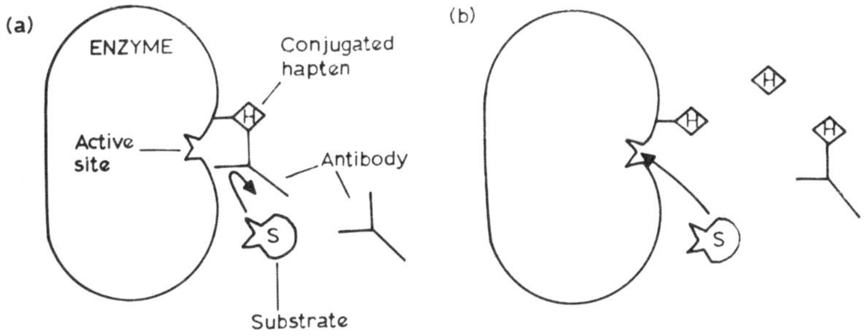

Fig. 3. Principle of homogeneous enzyme immunoassay. In this type of homogeneous assay, enzyme activity is inhibited when antibody is bound to the enzyme conjugate (a). The presence of hapten reduces the degree of antibody binding to conjugate and thus the degree of enzyme inhibition (b).

principle of these procedures and Table 2(b) the end product of some of these variants. Solid-phase-linked antigen or antigen-specific antibody is required for the binding and subsequent detection of antibody or antigen, respectively. In the latter case, an assay is performed by sequential addition of antigen and enzyme-labelled antibody to the solid-phase-linked antibody system with a washing step being performed after each addition. Addition of substrate enables quantification of the bound label.

Whichever procedure is used, the solid-phase reagent should ideally be present in sufficient excess to ensure complete immunoadsorption of the analyte in the standard or test sample. Similarly, an excess of all other reagents ensures complete sandwich formation for each analyte molecule. The process of sandwich formation obviously requires that the antigen be multivalent. The solid-phase antibody and labelled antibody must both be specific for the antigen but should be directed against different and spatially separated antigenic determinants. A typical calibration curve and assay protocol are shown in Fig. 4 and Table 3, respectively.

The enzyme-labelled species-specific anti-immunoglobulin is one of the more commonly used labels and forms the basis of the indirect ELISA (Table 2(b)). Although this procedure may involve an additional step in the assay, it obviates the need to label the analyte-specific antiserum or the analyte which may be in limited supply. Furthermore, the same label may be used in assays of a number of analytes depending upon the species of the specific antiserum. In order to avoid non-specific binding of label, it is important that the two analyte-specific antisera are raised in phylogenetically distant species, the labelled species-specific anti-

Table 2

Principles of the Sandwich Assay

(*a*) *Sandwich assay for detection of antigen*

}-Ab + Ag $\rightleftharpoons$ }-Ab...Ag $\rightleftharpoons$ }-Ab...Ag...Ab_1-Enz

(+ Ab_1-Enz)

(Kato *et al.*, 1975; Maiolini & Masseyeff, 1975; Ishikawa & Kato, 1978; Ishikawa *et al.*, 1982)

(*b*) *Final products of commonly employed types of sandwich assay*

(i) Detection of antibody. (Schmitz *et al.*, 1977)	$\rightleftharpoons$		$\rightleftharpoons$	}-Ag...Ab...Ag-Enz
(ii) Detection of antibody. (Schuurs & van Weemen, 1980)	$\rightleftharpoons$		$\rightleftharpoons$	}-Ag...Ab...Protein A-Enz
Indirect sandwich assays:				
(iii) Detection of antibody. (Engvall & Perlmann, 1972)	$\rightleftharpoons$		$\rightleftharpoons$	}-Ag...Ab_1...Ab_2-Enz
(iv) Detection of antigen. (Voller *et al.*, 1978)	$\rightleftharpoons$	$\rightleftharpoons$	$\rightleftharpoons$	}-Ab...Ag...Ab_1..Ab_2Enz

}- = Solid phase; Ab = Antigen-specific antibody; Ab_1 = antigen-specific antibody raised against different antigenic determinants to those of Ab; Ab_2 = antibody against immunoglobulin of the species in which Ab_1 was produced: in (iv) Ab_1 should therefore be raised in a different species than Ab; Protein A = a protein isolated from *S. aureus* which can interact specifically with the Fc region of most mammalian immunoglobulins.

immunoglobulin being raised against immunoglobulin of the species in which the second anti-analyte antibody (Ab_1 in Table 2(b)) was raised.

ENZYME-LABELLED REAGENTS

Choice of Enzyme

The choice of enzyme and its specific activity are primary considerations in relation to assay sensitivity since it will determine the size of signal at the assay end-point. Table 4 indicates important criteria for the selection of enzymes for use as labels. Few enzymes fulfill all these criteria but

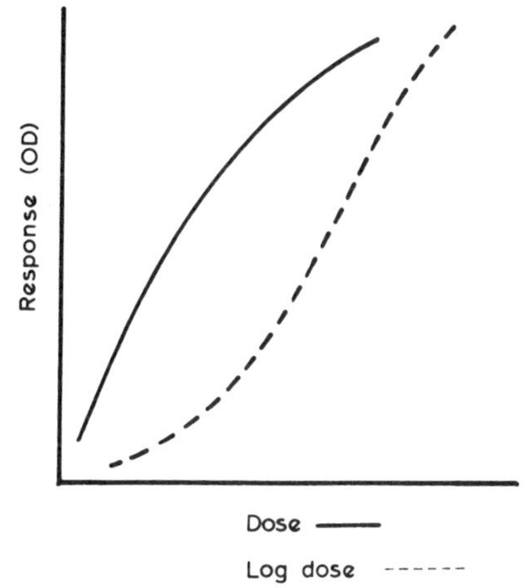

Fig. 4. Typical calibration curves for a sandwich assay.

Table 3
Typical Assay Protocol for a Microtitre Plate Sandwich Assay for the Determination of Antigen

1. Adsorb specific antibody onto wells of the microtitre plates (Voller *et al.*, 1976*a*), as Table 1. Store plates (containing buffer and a bacteriostat) at 4°C until required. Plastic film may be used to seal the plates.
2. Empty and wash wells with buffer prior to use.
3. Add sample or standard at a suitable dilution and incubate for 2 h at room temperature. Empty and wash the wells with buffer ($\times$ 3).
4. Add enzyme-labelled specific antibody and incubate for 2 h at room temperature. Empty and wash wells ($\times$ 3).
5. Add substrate solution, incubate for a suitable period and, if necessary (see Table 1), stop the reaction. Determine the absorbance.
6. Plot calibration curve and interpolate results.

Other types of sandwich assay (Table 2) are performed in a similar way. In the indirect assay for antigen a further incubation stage is required prior to addition of substrate.

enzymes such as β-galactosidase, alkaline phosphatase and peroxidase are most commonly used. All produce assays of approximately equivalent sensitivity and are available commercially in a suitably pure form and at reasonable cost. Given this, the prudent operator would perhaps

Table 4
Considerations for Selection of Enzyme

1. Purity of enzyme preparation.
2. Specific activity of enzyme.
3. Sensitivity of detection of products.
4. Absence of interfering factors or enzyme-like activity in test fluid.
5. Ease and speed of estimation of enzyme activity and the safety of the reagents used.
6. Availability and cost of enzyme.
7. Stability of the enzyme and its conjugate.

base selection on the ease of enzyme activity determination and the safety of the chemicals used. Peroxidase, for instance, is one of the least expensive enzymes and has many of the properties required for EIA, but the majority of chromogens used in its determination are mutagenic or carcinogenic: the formation of carcinogenic products is also likely whichever chromogen is used (Saunders *et al.*, 1964).

Although colorimetric end-points are preferred for most applications, β-galactosidase, alkaline phosphatase and peroxidase may also be assayed using substrates that form fluorescent products. This can increase the sensitivity of detection of enzyme by at least an order of magnitude (van der Waart & Schuurs, 1977; Ishikawa & Kato, 1978; Arakawa *et al.*, 1979; Ishikawa *et al.*, 1982).

Formation of Enzyme-Conjugates

Many potential users of EIA or ELISA are apprehensive of the chemistry involved in production of enzyme-labelled reagents. In ELISA, however, it is invariably possible to adopt procedures such as the indirect sandwich assay (Table 2(b)) in which commercially-available labels such as species-specific enzyme-labelled anti-immunoglobulin are used.

In general, enzyme-labelled forms of particular analytes or primary antibodies however, are not available commercially, and careful consideration should be given to the method of conjugate formation. Conjugates are required in which both the catalytic activity of the enzyme and immunoreactivity of the analyte or binding characteristics of the antibody are retained.

Conjugation of analyte or antibody with enzyme is generally achieved by covalent bonding through the lysine ε-amino groups of the enzyme or, where glycoprotein enzymes such as peroxidase are used, through carbohydrate residues. Methods of conjugation and their relative merits have been discussed by a number of authors (Schuurs & van Weemen, 1977; O'Sullivan *et al.*, 1979; Sauer *et al.*, 1982*a*) and will be described only briefly here.

Protein–Enzyme Conjugates

Since enzymes are themselves proteins containing reactive groups, the main problems in the production of enzyme–protein conjugates are related to the prevention of co-conjugation (polymerisation) of the enzyme or protein. Coupling reagents such as glutaraldehyde (Avrameas, 1969; Avrameas & Ternynck, 1971; Voller *et al.*, 1976*a*) and periodate oxidation (Nakane & Kawaoi, 1974) are still widely used and reported to provide conjugates of useful quality despite the probability of some degree of polymerisation and denaturation. It has been reported that formation of these undesirable by-products can be avoided using heterobifunctional reagents (Kitagawa & Aikawa, 1976; O'Sullivan *et al.*, 1978; Tae, 1983). Normally, incorporation of 1–2 enzyme molecules per antigen or antibody will be optimum since more extensive coupling may result in reduced enzyme activity or immunoreactivity.

Hapten–Enzyme Conjugates

Haptens may, in some instances, be coupled directly with enzymes, but usually it is necessary to form a derivative to provide a reactive group at a suitable location (Robinson *et al.*, 1975). Hapten–enzyme conjugates are commonly produced through formation of a peptide bond between carboxylic acid groups of the hapten and amino-groups on the enzyme or vice versa. Since enzymes contain both of these groups, methods which avoid polymerisation of the enzyme are preferred: mixed anhydride (Erlanger *et al.*, 1957) and *N*-hydroxysuccinimide ester (Anderson *et al.*, 1964) procedures have been used to covalently bind haptens or their derivatives to enzymes with minimal enzyme denaturation (Comoglio & Celada, 1976; Joyce *et al.*, 1978; Hosoda *et al.*, 1980; Sauer *et al.*, 1981). Direct carbodiimide condensation, on the other hand, will inevitably cause some degree of denaturation since the condensation reagent will necessarily come into contact with the enzyme with the result that the conjugate may need to be purified by fractionation or affinity chromatography (Dray *et al.*, 1975; Exley & Abuknesha, 1977). In theory, the coupling of one hapten per enzyme should be optimal, but in practice the location and orientation of the hapten on the enzyme may be important since the possibility of steric hindrance of antibody binding arises (Sauer *et al.*, 1981). It may, therefore, be necessary to attach several hapten molecules to each enzyme to ensure immunoreactivity: the optimum degree of incorporation may thus need to be determined experimentally and will depend on the enzyme and hapten employed.

Purification of Conjugates

It is essential that enzyme-labelled reagent is free from unlabelled analyte or antibody following conjugation to prevent a reduction in the sensitivity of the immunoassay. For both enzyme–hapten and enzyme–protein conjugates, this is most commonly achieved by gel filtration on the basis of molecular weight difference. The presence of unconjugated enzyme is rarely problematic since this is washed away with the unbound label during the immunoassay.

Conjugates can often be stored at 4°C in buffer containing preservatives such as sodium azide (β-galactosidase and alkaline phosphatase) or merthiolate (peroxidase). The preservative used should be carefully considered since some may inhibit enzyme activity.

SEPARATION OF FREE FROM ANTIBODY-BOUND LABEL

Numerous methods are employed for separation of free label from antibody-bound label (Table 5) although the most convenient means is through the attachment of one of the immunoreagents to a solid support. The antigen or antibody may be attached covalently to particulate material such as cellulose or agarose (Wide, 1981) or by passive adsorption to the inner walls of plastic tubes or microtitre plate wells (Engvall & Perlmann, 1971; Voller *et al.*, 1974). Separation of free and bound label during the immunoassay can be achieved by pouring-off or aspirating the free label: a particulate solid phase would obviously require to be sedimented before separation.

Immunoprecipitation (by the double antibody technique) has often been used in EIA, particularly in the performance of steroid hormone assays (Dray *et al.*, 1975; Comoglio & Celada, 1976; Joyce *et al.*, 1978). It offers the advantage of better availability of primary antibody to the other reagents, since all reagents remain in solution until the precipitating second antibody is added. It has been suggested that this may result in assays of increased sensitivity (Joyce *et al.*, 1978) although this has not been a general finding (Arakawa *et al.*, 1979; van Weemen *et al.*, 1979). This method does, however, require an extra incubation stage fol-

Table 5
Methods of Separating Bound from Free Label

1. Polyethylene glycol precipitation (Österman *et al.*, 1979).
2. Double antibody precipitation (Comoglio & Celada, 1976; Joyce *et al.*, 1978).
3. Double antibody solid-phase (Comoglio & Celada, 1976; Bosch *et al.*, 1978).
4. Solid-phase primary antibody:
   Agarose, cellulose, magnetisable cellulose (covalent binding: Engvall & Perlmann, 1971; Kato *et al.*, 1975; Guesdon & Avrameas, 1977; Joyce *et al.*, 1977; Turkes *et al.*, 1982).

   Polycarbonate-coated, polystyrene or nylon balls (covalent or passive binding: Miranda *et al.*, 1977; Smith & Gehle, 1977; Hendry & Herrman, 1980).

   Polystyrene test tubes (passive binding: Engvall & Perlmann, 1972; Stimson & Sinclair, 1974; Voller *et al.*, 1975).

   Polystyrene or polyvinyl microtitre plates (passive binding: Voller *et al.*, 1976*a*, 1976*b*; Sauer *et al.*, 1982*a*).

lowing addition of second antibody and so may be less convenient than solid phase EIA procedures. Again, separation is achieved following centrifugation.

EIA and ELISA procedures were considerably simplified with the adoption of disposable plastic microtitre plate systems (Voller *et al.*, 1976*b*) and these are worth particular consideration. The 125 mm × 85 mm plates are manufactured in polystyrene or polyvinyl and comprise 96 wells in a 8 × 12 format: dependent upon type, each well may contain a volume of up to 400µl. Antigen or antibody is passively adsorbed onto the inside of the wells and the plates stored until required. They may be sealed and kept at 4°C with buffer/preservative in the wells or dried and stored at room temperature in sealed dehydrated sachets, depending on the stability of the adsorbed material. During immunoassay, separation of free from bound label is performed simultaneously in all 96 wells by inversion and tapping dry over absorbent material. Over the past few years numerous automatic (through-the-well) colorimeters with microprocessor interface facilities have become available. These enable optical density determinations to be carried out without transfer of the samples into cuvettes. Since all 96 wells may be read in 1–1½ minutes it is not generally necessary to stop the enzyme reaction.

FACTORS INFLUENCING SENSITIVITY

The factors that influence sensitivity of EIA and ELISA parallel those in their radio-labelled analogues, although additional consideration must be given to factors which may modify enzyme activity and have been the subject of recent reviews (van der Waart & Schuurs, 1977; Sauer *et al.*, 1985).

The sensitivity of immunoassays is commonly defined as the smallest amount of analyte which gives rise to a response significantly different from zero at the 95% confidence level (limit of detection). This would be determined from the assay calibration curve and is defined as the mean value of the zero standard minus two standard deviations for EIA and for ELISA is the mean of the zero standard plus two standard deviations. When optimising conditions in order to improve sensitivity, the slope of the standard curve and the precision of determinations must be taken into account. A calibration curve of shallow slope may, for instance, be more capable of discriminating between similar concentrations of analyte than a steeper slope if precision of the former is greater (Ekins, 1983).

Influence of the Antiserum

The limit of detection of EIA is essentially determined, in accordance with the Law of Mass Action, by the avidity of binding of analyte to antibody: the more avid the antiserum the more sensitive will be the assay. In ELISA, the influence of avidity of binding is not so apparent since excess reagents are used and sensitivity will depend on the signal to noise ratio and, by implication, on the specific activity of the label (Ekins, 1978: Ishikawa & Kato, 1978).

The mass of antibody used in EIA or ELISA is determined by experiment through comparison of a range of antibody dilutions and conjugate concentrations in the presence or absence of clinically significant quantities of analyte (antibody dilution curves). In this way, the mass of reagents required to provide calibration curves over the range of clinical concentrations and with the required sensitivity can be ascertained. Objective assessment of the effect of alterations in method on assay performance may be aided by plotting precision profiles (Ekins, 1983).

The quantity of antigen required for detection of antibodies in ELISA would be similarly assessed by the preparation of dilution curves. For serological diagnosis of infectious diseases, e.g. *Trichinella spiralis* infection in pigs (see Bibliography index p. 20), however, only crude antigens are normally available. In these circumstances coating-concentrations of antigen for ELISA are normally determined by doing chequer-board titrations against positive and negative sera and subsequent tests performed using the antigen dilution which gave the greatest discrimination (Voller *et al.*, 1976*b*).

TIME AND TEMPERATURE OF INCUBATION

There are no hard and fast rules regarding optimum time and temperature for incubation, and where sensitivity is not critical, conditions are often adopted on the grounds of convenience. Short incubations may not be appropriate where large numbers of samples are involved since this can result in overlap of assays: 2–3h immunoreaction time is normally suitable. In ELISA, room temperature or 37°C incubations are common and can enable binding to approach completion within this period. EIA procedures are often carried out at 4°C however, since elevated temperature can drastically reduce assay sensitivity: where the antigen–antibody reaction involves a large enthalpy component, binding affinity will decrease with increased incubation temperature (Keane *et al.*, 1976). Enthalpy plays only a small part in the binding reaction of some antisera, however, and in these cases sensitivity may be relatively independent of incubation temperature (Malvano & Rolleri, 1975; Keane *et al.*, 1976). Such antisera should be identified at an early stage since they may enable incubations to be carried out at higher temperatures (e.g. 37°C or 40°C) and therefore for shorter periods without loss of sensitivity.

THE NATURE OF THE LINK BETWEEN HAPTEN AND ENZYME

RIAs of haptens are often performed using tritium-labelled hapten and the label is considered to be immunologically identical to the unlabelled hapten. In the case of enzyme-labelled hapten, however, the label may be bound with greater avidity than the native hapten because structurally it is very similar to the immunogen against which the antibodies were raised — a phenomenon referred to as 'bridge recognition'. This may result in a considerable loss in sensitivity (van Weemen & Schuurs, 1975) typified by a reduction in slope of the standard curve compared with tritium-based RIA using the same antiserum. Where such a loss in sensitivity is problematic, it has been found that changing the bridge linking the hapten to the enzyme and/or the site of attachment on the hapten can result in recovery of sensitivity (van Weemen & Schuurs, 1975; Exley & Abuknesha, 1977; Arakawa *et al.*, 1979; Hosoda *et al.*, 1980; Sauer *et al.*, 1982b).

NON-SPECIFIC EFFECTS

Non-specific adsorption of enzyme label, interference with specific binding and factors interfering with determination of enzyme activity have been reported to reduce assay performance, especially where relatively large volumes of undiluted or unextracted sample are assayed. Biological samples may, for instance, contain high concentrations of endogenous enzyme or substances which inhibit or enhance enzyme activity. In general, however, adequate wash procedures between assay stages can prevent such interference.

Non-specific adsorption of enzyme-label to the solid phase has been reported with solid-phase systems using passive adsorption of reagent to the solid phase (plastic tube and microtitre plate methods). This can normally be avoided by procedures which are now standard in most protocols: excess adsorption sites unoccupied by antibody may be blocked by addition of a high concentration of unrelated protein after antigen or antibody has been adsorbed. Alternatively, the inclusion of detergent or high molarity salt at the immuno-incubation stage can be used to inhibit non-specific binding (Engvall & Perlmann, 1971; Engvall *et al.*, 1971; Engvall & Ruoslahti, 1979; Saunders, 1979).

In steroid hormone EIA, as in RIA, the presence of steroid binding globulin in the plasma or serum of some species has been reported to interfere with the specific interaction of steroids and antibody. Specific displacement of steroid from the binding globulin has been achieved by reducing the pH at which the immunoreaction is performed (Riad-Fahmy *et al.*, 1981) or by prior heating of the sample (Ogihara *et al.*, 1977).

CONCLUSIONS

EIA and ELISA have been extended to many disciplines in the biological sciences and assays have been developed for a broad range of biochemical substances. These include serum proteins (Schuurs & van Weemen, 1977), hormones (Sauer *et al.*, 1982*a*), antibodies against infectious diseases (Voller *et al.*, 1978), viral antigens in infected plants (Voller *et al.*, 1976*c*; Clark & Adams, 1977), food constituents (Morgan *et al.*, 1983; chapters in this book and others listed in the Bibliography), antibiotics (Standefer & Saunders, 1978; Miura *et al.*, 1981) and herbicides (Niewola *et al.*, 1983). Heterogeneous EIA and ELISA are in limited use on a routine basis outside the sphere of infectious disease diagnosis although an increasing number of assays have been demonstrated to compare favourably with long established techniques (Oellerich, 1980; Schuurs & van Weemen, 1980).

The increasing demands made for more, and more varied, tests to be performed while staff numbers are reduced means that either tests will need to be circumvented by a greater degree of selectivity on the part of the person requesting the analysis or that they should be met by the use of simpler, more flexible and more rapid procedures. Microtitre plate and related tests hold much promise since they can be readily performed in a semi-automated form, using dispensing and washing devices and quantified using automatic plate readers. Under less favourable circumstances, qualitative and semi-quantitative results can be obtained using pipettes and wash bottles and the colour simply read by eye, comparisons being made with calibrators.

REFERENCES

ADDISON, G. M. & HALES, C. N. (1971) The immunoradiometric assay. In: *Radioimmunoassay Methods*, Kirkham, K. E. & Hunter, W. M. (eds), Churchill Livingstone, Edinburgh, UK, pp. 447–61.

ANDERSON, G. W., ZIMMERMAN, J. E. & CALLAHAN, F. M. (1964). The use of esters of *N*-hydroxysuccinimide in peptide synthesis. *Journal of the American Chemical Society*, **86**, 1839–42.

ARAKAWA, H., MAEDA, M. & TSUJI, A. (1979) Chemiluminescence enzymeimmunoassay of cortisol using peroxidase as label. *Analytical Biochemistry*, **97**, 248–54.

AVRAMEAS, S. (1969) Coupling of enzymes to proteins with glutaraldehyde. Use of the conjugates for the detection of antigens and antibodies. *Immunochemistry*, **6**, 43–52.

AVRAMEAS, S. & TERNYNCK, T. (1971) Peroxidase labelled antibody and Fab conjugates with enhanced intracellular penetration. *Immunochemistry*, **8**, 1175–9.

BERSON, S. A. & YALOW, R. S. (1959) Quantitative aspects of the reaction between insulin and insulin-binding antibody. *Journal of Clinical Investigation*, **38**, 1996–2016.

BOGUSLASKI, R. C. & LI, T. M. (1982) Homogeneous immunoassays. *Applied Biochemistry & Technology*; **7** (5), 401–14.

BOSCH, A. M. G., DIJKHUIZEN, D. M., SCHUURS, A. H. W. M. & VAN WEEMEN, B. K. (1978) Enzyme immunoassay for total oestrogens in pregnancy plasma or serum. *Clinica Chimica Acta*, **89**, 59–70.

CLARK, M. F. & ADAMS, A. N. (1977) Characteristics of the microplate method of enzyme-linked immunosorbent assay for the detection of plant viruses. *Journal of General Virology*, **34**, 475–83.

COMOGLIO, S. & CELADA, F. (1976) An immuno-enzymatic assay of cortisol using *E. coli* β-galactosidase as label. *Journal of Immunological Methods*, **10**, 161–70.

DRAY, F., ANDRIEU, J. M. & RENAUD, F. (1975) Enzyme immunoassay of progesterone at the picogram level using β-galactosidase as label. *Biochimica Biophysica Acta*, **403**, 131–8.

EKINS, R. P. (1978) The future of development of immunoassay. In: *Radioimmunoassay and Related Procedures in Medicine*, Vol. 1, Proceedings of an International Symposium, 1977, IAEA, Vienna, pp. 241–75.

EKINS, R. P. (1983) The precision profile: its use in assay design, assessment and quality control. In: *Immunoassays for Clinical Chemistry*, 2nd edn, Hunter, W. M. & Corrie, J. E. T. (eds), Churchill Livingstone, Edinburgh, UK, pp. 76–104.

ENGVALL, E. & PERLMANN, P. (1971) Enzyme-linked immunosorbent assay (ELISA). Quantitative assay of immunoglobulin G. *Immunochemistry*, **8**, 871–4.

ENGVALL, E. & PERLMANN, P. (1972) Enzyme-linked immunosorbent assay, ELISA. III. Quantitation of specific antibodies by enzyme-labeled anti-immunoglobulin in antigen-coated tubes. *Journal of Immunology*, **109**, 129–35.

ENGVALL, E. & RUOSLAHTI, E. (1979) Principles of ELISA and recent applications to the study of molecular interactions. In: *Immunoassays in the Clinical Laboratory*, Nakamura, R. M., Dito, W. R. & Tucker, E. S. (eds), Alan R. Liss, New York, pp. 89–97.

ENGVALL, E., JONSSON, K. & PERLMANN, P. (1971) Enzyme-linked immunosorbent assay II. Quantitative assay of protein antigen, immunoglobulin G, by means of enzyme-labelled antigen and antibody-coated tubes. *Biochimica Biophysica Acta*, **251**, 427–34.

ERLANGER, B. F., BOREK, F., BEISER, S. M. & LIEBERMAN, S. (1957) Steroid–protein conjugates. 1. Preparation and characterization of conjugates of bovine serum albumin with testosterone and with cortisone. *Journal of Biological Chemistry*, **228**, 713–27.

EXLEY, D. & ABUKNESHA, R. (1977). The preparation and purification of a β-D-galactosidase–oestradiol-17β conjugate for enzyme immunoassay. *FEBS Letters*, **79**, 301–4.

GUESDON, J-L. & AVRAMEAS, S. (1977) Magnetic solid phase enzyme-immunoassay. *Immunochemistry*, **14**, 443–7.

HENDRY, R. M. & HERRMAN, J. E. (1980) Immobilization of antibodies on nylon for use in enzyme-linked immunoassay. *Journal of Immunological Methods*, **35**, 285–96.

HOSODA, H., YOSHIDA, H., SAKAI, Y., MIYAIRI, S. & NAMBARA, T. (1980) Sensitivity and specificity in enzymeimmunoassay of testosterone. *Chemical and Pharmaceutical Bulletin*, **28**, 3035–40.

ISHIKAWA, E. & KATO, K. (1978) Ultrasensitive enzyme immunoassays. *Scandinavian Journal of Immunology*, **8** (Suppl. 7), 43–55.

ISHIKAWA, E., IMAGAWA, M., YOSHITAKE, S., NIITSU, Y., URUSHIZAKI, I., INADA, M., IMURA, H., KANAZAWA, R., TACHIBANA, S., NAKAZAWA, N. & OGAWA, H. (1982) Major factors limiting sensitivity of sandwich enzymeimmunoassay for ferritin, immunoglobulin E, and thyroid-stimulating hormone. *Annals of Clinical Biochemistry*, **19**, 379–84.

JOYCE, B. G., READ, G. F. & FAHMY, D. R. (1977) Specific enzymeimmunoassay for progesterone in human plasma. *Steroids*, **29**, 761–70.

JOYCE, B. G., WILSON, D. W., READ, G. F. & RIAD-FAHMY, D. (1978) An improved enzyme immunoassay for progesterone in human plasma. *Clinical Chemistry*, **24**, 2099–102.

KATO, K., HAMAGUCHI, Y., FUKUI, H. & ISHIKAWA, E. (1975) Coupling Fab[1] fragment of rabbit anti-human IgG antibody to β-D-galactosidase and a highly sensitive immunoassay of human IgG. *FEBS Letters*, **56**, 370–2.

KEANE, P. M., WALKER, W. H. C., GAULDIE, J. & ABRAHAM, G. E. (1976) Thermodynamic aspects of some radioassays. *Clinical Chemistry*, **22**, 70–3.

KITAGAWA, T. & AIKAWA, T. (1976) Enzyme coupled immunoassay of insulin using a novel coupling reagent. *Journal of Biochemistry* (*Tokyo*), **79**, 233–6.

MAIOLINI, R. & MASSEYEFF, R. (1975) A sandwich method of enzymeimmunoassay. I. Application to rat and human alpha-fetoprotein. *Journal of Immunological Methods*, **8**, 223–34.

MALVANO, R. & ROLLERI, E. (1975) Methodological aspects of steroid radioimmunoassay. In: *Radioimmunoassay of Steroid Hormones*, Gupta, D. (ed.), Verlag Chemie, Weinheim, FRG, pp. 27–53.

MILES, L. E. M. & HALES, C. N, (1968) Labelled antibodies and immunological assay systems. *Nature*, **219**, 186–9.

MIRANDA, Q. R., BAILEY, G. D., FRASER, A. S. & TENOSO, H. J. (1977) Solid-phase enzyme immunoassay for Herpes-simplex virus. *Journal of Infectious Diseases Supplement*, **136**, S304–S310.

MIURA, T., KOUNO, H. & KITAGAWA, T. (1981) Detection of residual penicillin in milk by sensitive enzyme immunoassay. *Journal of Pharmacobio-Dynamics*, **4**, 706–10.

MORGAN, M. R. A., MCNERNEY, R., MATTHEW, J. A., COXON, D. T. & CHAN, H. W-S. (1983) An enzyme-linked immunosorbent assay for total glycoalkaloids in potato tubers. *Journal of the Science of Food and Agriculture*, **34**, 593–8.

NAKANE, P. K. & KAWAOI, A. (1974) Peroxidase-labelled antibody. A new method of conjugation. *Journal of Histochemistry Cytochemistry*, **22**, 1084–91.

NIEWOLA, Z., WALSH, S. T. & DAVIES, G. E. (1983) Enzyme-linked immunosorbent assay (ELISA) for paraquat. *International Journal of Immunopharmacology*, **5**, 211–18.

OELLERICH, M. (1980) Enzyme immunoassays in clinical chemistry: present status and trends. *Journal of Clinical Chemistry and Clinical Biochemistry*, **18**, 197–208.

OGIHARA, T., MIYAI, K., NISHI, K., ISHIBASHI, K. & KUMAHARA, Y. (1977) Enzyme-labeled immunoassay for plasma cortisol. *Journal of Clinical Endocrinology and Metabolism*, **44**, 91–5.

OLIVER, G. C. (JR), PARKER, B. M., BRASFIELD, D. L. & PARKER, C. W. (1968) The measurement of digitoxin in human serum by radioimmunoassay. *Journal of Clinical Investigation*, **47**, 1035–42.

ÖSTERMAN, T. M., JUNTUNEN, K. O. & GOTHONI, G. D. (1979) Enzyme immunoassay of oestrogen-like substances in plasma, with polyethylene glycol as precipitant. *Clinical Chemistry*, **25**, 716–18.

O'SULLIVAN, M. J., GNEMMI, E., MORRIS, D., CHIEREGATTI, G., SIMMONS, M., SIMMONDS, A. D., BRIDGES, J. W. & MARKS, V. (1978) A simple method for the preparation of enzyme–antibody conjugates. *FEBS Letters*, **95**, 311–13.

O'SULLIVAN, M. J., BRIDGES, J. W. & MARKS, V. (1979) Enzyme immunoassay: A review. *Annals of Clinical Biochemistry*, **16**, 221–39.

RIAD-FAHMY, D., READ, G. F., JOYCE, B. G. & WALKER, R. F. (1981) Steroid immunoassays in endocrinology. In: *Immunoassays for the 80's*, Voller, A., Bartlett, A. & Bidwell, D. (eds), MTP Press Ltd, Lancaster, UK, pp. 205–61.

ROBINSON, J. D., MORRIS, B. A., PIALL, E. M., AHERNE, G. W. & MARKS, V. (1975) The use of rats in the screening of drug–protein conjugates for immunoreactivity. In *Radioimmunoassay in Clinical Biochemistry*, Pasternak, C. A. (ed.), Heyden, London, pp. 101–11.

SAUER, M. J., FOULKES, J. A. & COOKSON, A. D. (1981) Direct enzymeimmunoassay of progesterone in bovine milk. *Steroids*, **38**, 45–53.

SAUER, M. J., COOKSON, A. D., MACDONALD, B. J. & FOULKES, J. A. (1982*a*) The use of enzymeimmunoassay for the measurement of hormones with particular reference to the determination of progesterone in unextracted whole milk. In *The ELISA: Enzyme-linked Immunosorbent Assay in Veterinary Research and Diagnosis*, Wardley, R. C. & Crowther, J. R. (eds), Nijhoff, The Hague, pp. 271–301.

SAUER, M. J., FOULKES, J. A. & O'NEILL, P. M. (1982*b*) Use of microtitre plate EIA for direct determination of progesterone in whole milk: application of heterologous systems for improved sensitivity. *British Veterinary Journal*, **138**, 522–31.

SAUER, M. J., FOULKES, J. A. & MORRIS, B. A. (1985) Factors affecting sensitivity of heterogeneous enzyme immunoassay. In: *Immunoassays in Veterinary Practice*, Morris, B. A. & Bolton, A. E. (eds), Ellis Horwood, Chichester, UK (in press).

SAUNDERS, G. C. (1979) The art of solid-phase enzyme immunoassay including selected protocols. In: *Immunoassays in the Clinical Laboratory*, Nakamura, R. M., Dito, W. R. & Tucker, E. S. (eds), Alan R. Liss, New York, pp. 99–118.

SAUNDERS, G. C., HOLMES-SIEDLE, A. G. & STARK, B. P. (eds) (1964). Peroxidase, Butterworths Co. Ltd, London.

SCHALL, R. F. (JR) & TENOSO, H. J. (1981) Alternatives to radioimmunoassay: labels and methods. *Clinical Chemistry*, **27**, 1157–64.

SCHMITZ, H., DOERR, H-W., KAMPA, D. & VOGT, A. (1977) Solid-phase enzyme immunoassay for immunoglobulin M antibodies to cytomegalovirus. *Journal of Clinical Microbiology*, **5**, 629–34.

SCHUURS, A. H. W. M. & VAN WEEMEN, B. K. (1977) Enzyme-immunoassay. *Clinica Chimica Acta*, **81**, 1–40.

SCHUURS, A. H. W. M. & VAN WEEMEN, B. K. (1980) Enzyme-immunoassay: A powerful analytic tool. *Journal of Immunoassay*, **1**, 229–49.

SMITH, K. O. & GEHLE, W. D. (1977) Magnetic transfer devices for use in solid-phase radioimmunoassay and enzyme-linked immunosorbent assays. *Journal of Infectious Diseases, Supplement*, **136**, S329–S336.

STANDEFER, J. C. & SAUNDERS, G. C. (1978) Enzyme immunoassay for gentamicin. *Clinical Chemistry*, **24**, 1903–7.

STIMSON, W. H. & SINCLAIR, J. M. (1974) An immunoassay for pregnancy-associated α-macroglobulin using antibody–enzyme conjugates. *FEBS Letters*, **47**, 190–2.

TAE, H. JI. (1983) Bifunctional reagents. *Methods in Enzymology*, **91**, 580–609.

TURKES, A., READ, G. F. & RIAD-FAHMY, D. (1982) A simple high-throughput enzymeimmunoassay for norethisterone (norethindrone). *Contraception*, **252**, 505–14.

VAN DER WAART, M. & SCHUURS, A. H. W. M. (1977) The sensitivity of enzyme-immunoassay. Some facts and figures. *Bulletin-Schweizerische Gesellschaft für Klinische Chemie*, (4), 9–16.

VAN WEEMEN, B. K. & SCHUURS, A. H. W. M. (1971) Immunoassay using antigen–enzyme conjugates. *FEBS Letters*, **15**, 232–6.

VAN WEEMEN, B. K. & SCHUURS, A. H. W. M. (1975) The influence of heterologous combinations of antiserum and enzyme-labelled estrogen on the characteristics of estrogen enzyme-immunoassays. *Immunochemistry*, **12**, 667–70.

VAN WEEMEN, B. K., BOSCH, A. M. G., DAWSON, E. C. & SCHUURS, A. H. W. M. (1979) Enzymeimmunoassay of steroids: possibilities and pitfalls. *J. Steroid Biochemistry*, **11**, 147–51.

VOLLER, A., BIDWELL, D., HULDT, G. & ENGVALL, E. (1974) A microplate method of enzyme-linked immunosorbent assay and its application to malaria. *Bulletin of the World Health Organisation*, **51**, 209–11.

VOLLER, A., HULDT, G., THORS, C. & ENGVALL, E. (1975) New serological tests for malaria antibodies. *British Medical Journal*, **1**, 659–61.

VOLLER, A., BIDWELL, D. E. & BARTLETT, A. (1976*a*). Enzymeimmunoassay for diagnostic medicine. Theory and practice. *Bulletin of the World Health Organisation*, **53**, 55–65.

VOLLER,. A., BIDWELL, D. E. & BARTLETT, A. (1976*b*) Microplate enzyme immunoassays for the immunodiagnosis of virus infections. In: *Manual of Clinical Immunology*, Rose, N. R. & Friedman, H. (eds), American Society for Microbiology, Washington, DC, pp. 506–12.

VOLLER, A., BARTLETT, A., BIDWELL, D. E., CLARK, M. F. & ADAMS, A. N. (1976*c*) The detection of viruses by enzyme-linked immunosorbent assay (ELISA). *Journal of General Virology*, **33**, 165–7.

VOLLER, A., BARTLETT, A. & BIDWELL, D. E. (1978) Enzyme immunoassays with special reference to ELISA techniques. *Journal of Clinical Pathology*, **31**, 507–20.

WIDE, L. (1981) Use of particulate immunosorbents in radioimmunoassay. *Methods in Enzymology*, **73**, 203–24.

4

Alternative Labels in Non-isotopic Immunoassay

G. W. AHERNE

Division of Clinical Biochemistry, Department of Biochemistry, University of Surrey, Guildford, UK

INTRODUCTION

Radioimmunoassay (RIA) techniques were first developed over twenty years ago and have now been widely applied in many fields of study, especially in clinical chemistry. An immense range of substances has been quantified using the technique, the success of which lies in the fact that a high degree of both specificity and sensitivity can be achieved. Radioimmunoassays do however suffer from several disadvantages. Radioisotopically labelled antigens have limited stability. Whereas ^{125}I-labelled compounds have short shelf lives, mainly due to the half-life of the isotope (60 days), antigens labelled with ^{3}H or ^{14}C are often chemically unstable and also need replacing at intervals of 6–12 months. The cost of radiolabelled antigens is therefore relatively high and the capital cost of radioactivity counting instruments as well as of scintillation cocktails should also be borne in mind. Although biological samples do not interfere with the radioactive measurements, the signal cannot be modified and therefore a separation phase is required. Automation of RIAs has therefore presented some difficulties.

The need to perform frequent radioiodinations is a cause of concern as a potential hazard to health and the use of radioactivity in a particular establishment requires special authorisation. There are also strict regulations in many countries governing the disposal of radioactivity.

A number of alternatives to radioactive labels have been investigated, and if they are to be widely applied, should have attributes of stability, with a greatly reduced potential health risk, should retain the potential of sensitivity and show little or no interference from sample components, should be relatively inexpensive, and it should be possible to measure the end-point with equipment which is readily available and inexpensive.

Immunoassays using stable free radicals (Esser, 1980), particles (Leuvering *et al.*, 1980) or viruses (Andrieu *et al.*, 1975) have been described but have not been widely used. Enzymes and coenzymes have been the most successfully used alternatives to radiolabels (Voller *et al.*, 1978) and their use has been reviewed in another chapter. Recently, a great deal of interest has been shown in the use of fluorescent and luminescent labels. The practical advantages of using these non-isotopic labels as well as some of the problems encountered in their use will be discussed here. Future developments in both fluoroimmunoassay (FIA) and luminoimmunoassay (LIA) will also be considered.

FLUOROIMMUNOASSAYS (FIA)

Several fluoresent molecules have been used for labelling both antigens and antibodies. The labels have proved to be stable over a long period of time and can be produced inexpensively. The fluorescent signal of the assay end-point can be measured quickly and simply using readily available and relatively inexpensive equipment.

Ideally molecules which are used as fluorescent labels should have a high fluorescent intensity and the fluorescence should be distinguishable from background. Fluorescence-labelled antigens or antibodies should retain much of the quantum efficiency of the fluorophore and at the same time retain immunoreactivity. Fluoroscein has been the most commonly used label although other fluorophores, e.g. rhodamines, umbelliferones and rare earth metal chelates, have also been used.

There are, however, a number of problems associated with the use of fluorophores in biological systems. These include light scattering effects, caused by macromolecular components of the sample, e.g. lipids, which produce an apparent increase in fluorescence, and inner filter effects, caused by absorption of part of either the excitation or emission beam by other chromophores such as haemoglobin. Quenching of fluorescence can be caused by a number of factors in the environment of a fluorophore, including protein binding either non-specifically to albumin or to specific antibodies. The main problem, though, is due to the presence of endogenous fluorophores in the samples. Compounds such as bilirubin cause a high background fluorescence in the same region as most fluorescent probes at the same excitation wavelength. Some of these problems associated with the measurement of fluorescence in biological samples can be minimised by the use of fluorometers incorporating high quality filters and monochromators.

Types of FIA

Separation Assays

Many FIA techniques are, in practical terms, directly comparable to RIA techniques, except that the radiolabel is replaced by a fluorescent label. Following phase separation the fluorescence is measured in either the antibody-bound fraction or the free (unbound) fraction. Background and interference problems can be decreased using separation assays, especially if a washing procedure can be incorporated. Solid-phase techniques have been widely applied to FIAs (Burgett *et al.*, 1977; Curry *et al.*, 1979) although some plastics can contribute to the background fluorescence and scattering effects. Immunofluorometric assays (IFMA) using antibodies labelled with fluorescent probes have also been described. These include assays for haptens (Ekeke *et al.*, 1979) as well as for proteins (Aalberse, 1973) where a 'sandwich' type assay was used.

Non-separation Assays

One major advantage compared to the use of radiolabels in immunoassay is that they offer the possibility of developing non-separation or homogeneous assays. Several types of non-separation assays have been described but their use has been generally limited and they often suffer from low sensitivity due to background fluorescence.

Quenching FIAs

These depend upon the quenching of the fluorescence of a labelled antigen as a result of binding to specific antibody. The assays are simple to perform and offer the possibility of automation. Direct quenching FIAs are applicable only to the assay of haptens (Shaw *et al.*, 1977; Broughton & Frazier, 1978). Indirect quenching FIA techniques have therefore been applied to the measurement of proteins (Zuk *et al.*, 1979) and depend upon the steric hindrance of the antibody in an antigen–antibody complex preventing the attachment of a quenching antibody (e.g. anti-fluoroscein). In another type of quenching assay (Ullman *et al.*, 1976), the antigen is labelled with one fluorophore, e.g. fluoroscein, and the antibody with another, e.g. rhodamine. The latter absorbs the emission from the fluoroscein when antigen–antibody complexes are formed.

Enhancement FIAs
Unusually, a FIA based on enhancement of the fluorescence of a labelled antigen when it is bound to antibody has been described (Smith, 1977). The fluorescence of fluoroscein bound to thyroxine was thought to be quenched by the iodine of thyroxine and upon antibody binding the fluorescence was increased four-fold.

Release FIA
In this type of assay (Burd *et al.*, 1977), a non-fluorescent probe, such as umbelliferyl-β-galactoside, which acts both as an enzyme substrate and as a precursor of fluorescence, is used. Under the action of a suitable enzyme, β-galactosidase, the non-fluorescent precursor-substrate is hydrolysed to its fluorescent product. However, binding of the non-fluorescent label by antibody prevents enzymatic hydrolysis and thereby the release of fluorescence. Unlabelled antigen competing with the label for antibody binding would cause an increase in fluorescence proportional to the amount of unlabelled antigen added. Such assays have been limited to the measurement of haptens and are ideally suited to therapeutic drug monitoring although the range of commercial kits available is for only a small range of drugs.

Polarisation FIA
In this non-separation type of assay, a sample containing antigen labelled with a fluorophore is excited with polarised light and the extent of polarisation of the emitted light is measured. Small labelled molecules, with fast random rotation, will exhibit a low signal. Antibody–antigen complexes, with increased size and reduced random rotation, will exhibit an enhanced signal. The technique has been applied to a number of analytes (Tengerdy, 1967; Watson *et al.*, 1976; Urios *et al.*, 1978) and several kits are available commercially. Although the assays are simple and quick to perform, their use depends on the availability of polarisation fluorimeters which are very expensive. Other disadvantages include the low sensitivity achieved, a limited dynamic range and precise optimisation requirements, as well as the lack of applicability to antigens with a molecular weight of greater than 20 000.

Future Developments
In spite of their attributes of speed, simplicity and ease of end-point measurement, FIA techniques have been limited in some clinical applications by disappointingly low sensitivity compared to RIA. This low sensitivity (10^{-7}–10^{-8} M) will not necessarily be so limiting to food analysis. Although many types of assay have been described, very few of these have been used in laboratories other than those directly involved in their development. The most important use of FIA on a large scale has been in the field of therapeutic drug monitoring.

Attempts to improve assay sensitivity continue. One exciting development, which remains to be investigated further, is the use of time-resolved fluorimetry. Chelates of the rare earth metals have a long lived fluorescence and high quantum yield. Interference from background fluorescence, which is short lived, can be overcome by measuring the fluorescence caused by a short excitation pulse after a certain time delay (during which time background fluorescence is totally reduced). In one assay (Pettersson *et al.*, 1983) a europium chelate was used to label antibodies to human choriogonadotropin. These were used in a rapid, solid-phase sandwich assay for the hormone in serum over a wide measurement range. Time-resolved FIA is an attractive development in immunoassay, potentially more sensitive than RIA, but its wide applicability remains to be proved in the next few years. The use of phosphorescent probes is also under investigation as an alternative to radioactivity in immunoassay.

LUMINOIMMUNOASSAYS (LIA)

Several new developments in analytical chemistry and clinical chemistry have involved the use of chemiluminescent molecules (Whitehead *et al.*, 1979) and chemiluminescence dependent methods of analysis are already familiar to the food scientist. Luminescent labels in immunoassay have many potential advantages over the use of radiolabels, e.g. stability, ease of preparation. Luminescence can be measured easily in a few seconds using relatively simple light measuring instruments. Simple as well as semi-automated/automated luminometers are now commercially available, but are not primarily designed for immunoassay purposes. The measurement of luminescence can be made over a theoretical background of zero, although errors can occur due to both physical and chemical effects. Although luminescence measurements are insensitive to turbidity, they can be affected by light absorbing or fluorescent molecules present in the sample. Temperature and pH are also factors which should be standardised.

Chemiluminescence is produced in certain chemical reactions in which the energy released during the reaction is sufficient to produce light quanta (Fig. 1(a)). Most chemiluminescent reactions are oxidation reactions. Molecules in a low energy ground state obtain energy from oxidation and, as energy is released in the form of chemiluminescence, fall from a high energy to a low energy state again. Bioluminescence occurs in many living organisms, the oxidation step being catalysed by a group of enzymes, the luciferases (Fig. 1(b)). The substrates for the luciferases are known as luciferins.

Both firefly and bacterial luciferase have been used to label antigens in immunoassay procedures (Olsson *et al.*, 1979*b*; Wannlund *et al.*, 1980). In theory, the substrate or any of the co-factors, e.g. ATP, NAD, required for the luciferase reaction can be used for labelling purposes (Carrico *et al.*, 1976; Schroeder *et al.*, 1976).

The most extensively studied luminescent compound is luminol (5-amino-2, 3-dihydrophthalazine-1, 4-dione). Chemiluminescence is produced when luminol is oxidised by hydrogen peroxide or other peroxides. The reaction proceeds much faster in alkaline conditions and in the presence of a catalyst such as metal containing complexes or enzymes containing haem, e.g. microperoxidase. The most commonly used oxidation system for luminol is hydrogen peroxide and microperoxidase, the detection limit for which is 1 pmol/litre (Schroeder & Yeager, 1978). Unfortunately the luminescent activity of luminol can be partially or totally abolished in conjugates with antigens or antibodies. Isoluminol, which in itself, is less efficient at producing luminescence, has been used as an alternative to luminol. Conjugation does not result in loss of luminescence because of the more favourable position of the amino group. Several derivatives of isoluminol (Fig. 2) containing alkyl side chains of varying length are now available, e.g. ABEI (*N*-(4-aminobutyl)-*N*-ethyl isoluminol), AHEI (*N*-(6-aminohexyl)-*N*-ethyl isoluminol) and ABEI-H (*N*-(4-aminobutyl)-*N*-ethyl isoluminol hemisuccinamide), which facilitate conjugation by providing different functional and spacer groups. Chemical reactions, already familiar to immunoassayists (mixed anhydride and carbodiimide condensations, diazotisation, etc.) can be used to prepare luminescent labels for haptens, proteins and antibodies. Peroxidase or microperoxidase-labelled conjugates, which are established

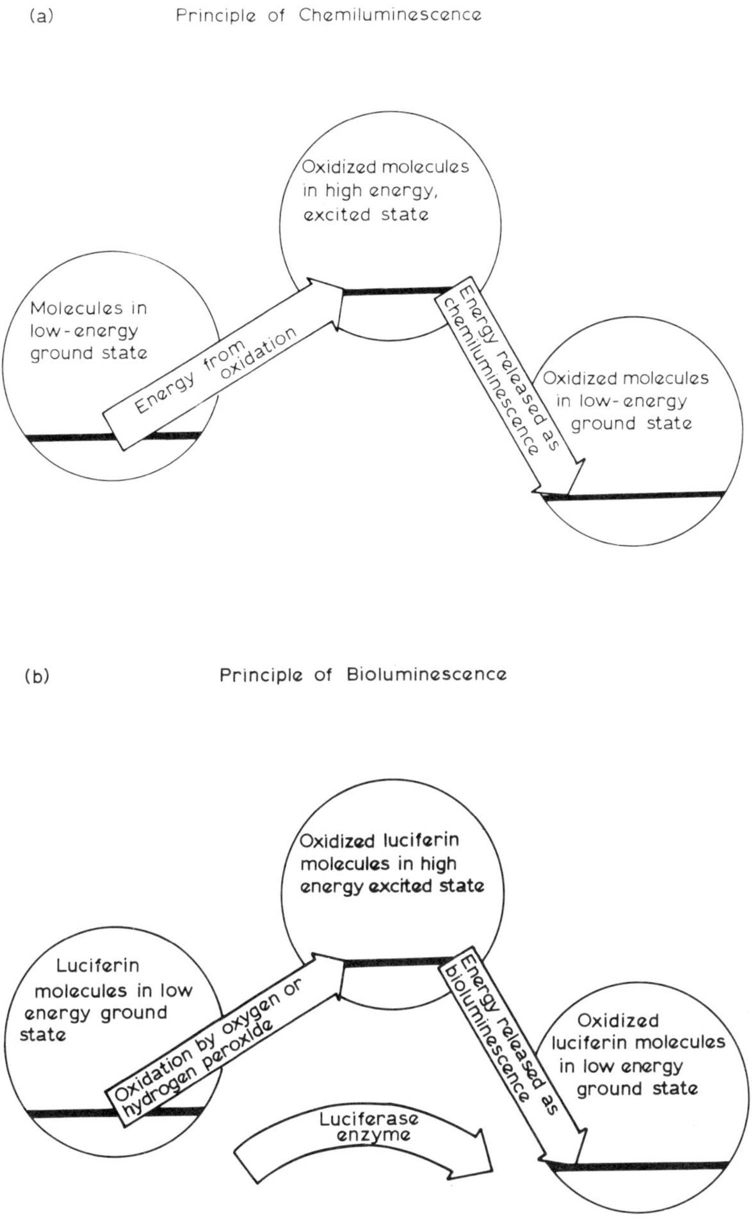

Fig. 1. The principle of (a) chemiluminescence and (b) bioluminescence.

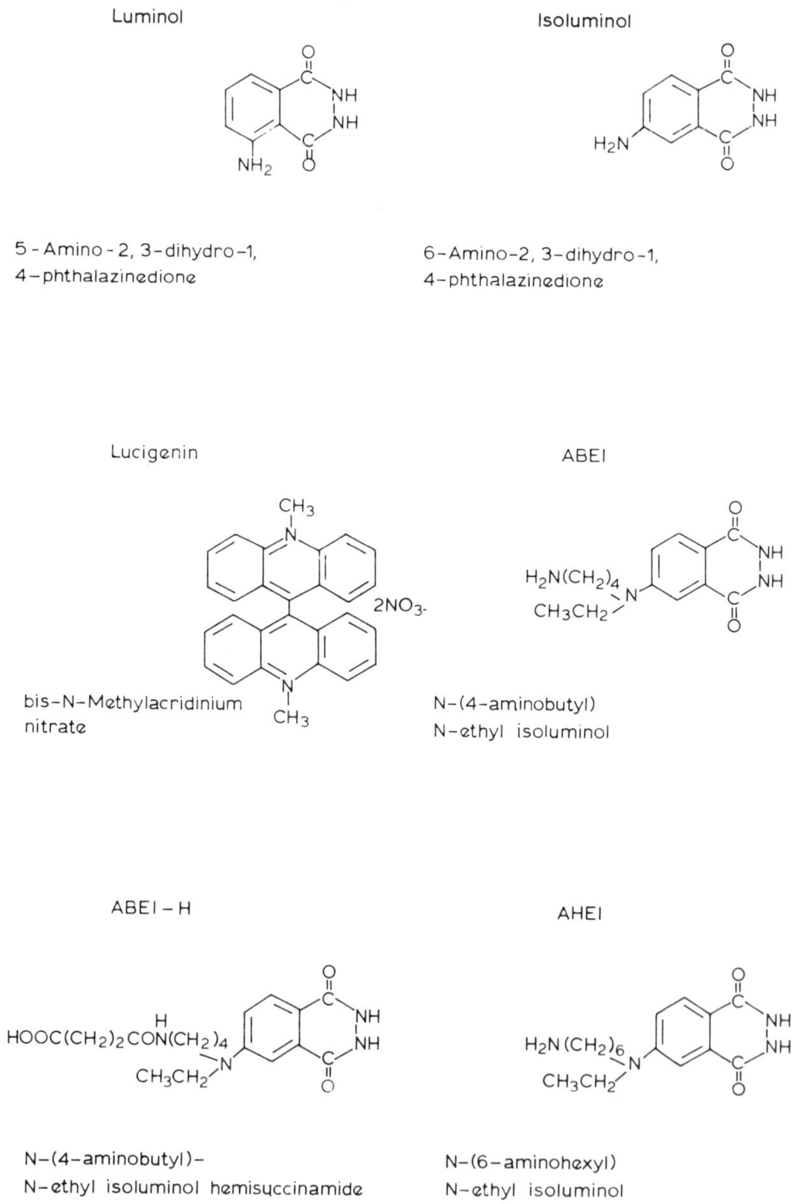

Fig. 2. Structural formulae of some luminescent derivatives used in immunoassays.

immunoassay reagents, are active in the luminol system (Arakawa *et al.*, 1979).

Aryl acridinium esters have also been used for labelling purposes in immunoassay (Woodhead *et al.*, 1981). Luminescence is produced on oxidation by peroxide in basic solutions containing metal ion catalysts. Although the acridinium salts can be detected under milder conditions and with greater sensitivity than luminol, much less work in immunoassay systems has been described.

Types of LIA

Luminescent labels can be easily incorporated into established RIA techniques with similar or improved sensitivity and short measurement times. One disadvantage is that following phase separation, a period of incubation (approximately 1 h) at alkaline pH is required for optimal light production. The conditions of this incubation should be carefully optimised for each assay system. Several solid-phase assays have been reported which utilise luminescent labels (Olsson *et al.*, 1979*a*; Barnard *et al.*, 1981; Lindstrom *et al.*, 1982; Weerasekera *et al.*, 1983) results from which correlate well with those from RIA. Immunochemiluminometric assays (ICMA) have also been described for the measurement of protein hormones (Simpson *et al.*, 1979) that are as sensitive as the equivalent RIA.

In contrast to RIA, it is possible to establish homogeneous or non-separation assays using luminescent probes (Kohen *et al.*, 1979). These assays depend upon the enhancement of light intensity when the labelled antigen is bound to specific antibody and this may be due to the presence of a more favourable hydrophobic environment for the emission of light. In one assay (Kohen *et al.*, 1980), the light emission of a cortisol-isoluminol derivative was delayed when bound to specific cortisol antibodies. A sensitive homogeneous assay for cyclic nucleotides has been established which is based on energy transfer when a chemiluminescent labelled antigen is bound to a fluorescent-labelled antibody (Campbell & Patel, 1983). It has been demonstrated that homogeneous LIAs are as sensitive or more so than conventional RIA but are subject to interference from biological samples. For this reason, solid-phase assays may be preferable.

Future Developments

The use of chemiluminescence labels in immunoassays has many attractions and their use is presently being investigated in a number of laboratories including the author's. Purified luminol and acridinium derivatives are available commercially as are some labelled antigens. As time progresses a large range of labels should become available. Luminometer design is likely to improve and to take more account of the needs of the immunoassayist. At the moment, luminescence is enhanced using an incubation with sodium hydroxide and studies are required to define conditions where this incubation period can be shortened or dispensed with.

SUMMARY

Much effort has been devoted to finding alternatives to the use of radioactivity in immunoassays. However, radioimmunoassays remain the most commonly used type of immunoassay, although enzyme immunoassays have made a large and increasing impact for certain applications. FIA and LIA have tremendous potential as alternatives to RIA but their widespread acceptance will depend upon a number of factors. In laboratories and countries where the use of radioactivity is undesirable or not permitted by legislation, non-isotopic immunoassays are essential. Ultimately cost, availability of reagents and equipment, type of analyte under consideration and sensitivity required will be the main determinants of which type of immunoassay is carried out in a particular laboratory.

REFERENCES

AALBERSE, R. C. (1973) Quantitative fluoroimmunoassay. *Clinica Chimica Acta*, **48**, 109–13.

ANDRIEU, J. M., MANAS, S. & DRAY, F. (1975) Viroimmunoassay of steroids: Method and principles. In: *Steroid Immunoassay — Proceedings of the Fifth Tenovus Workshop*, Cardiff, April 1974, Cameron, E. H. D., Hiller, S. G. & Griffiths, K. (eds), Alpha Omega Publishing, Cardiff, UK, pp. 189–98.

ARAKAWA, H., MAEDA, M. & TSUJI, A. (1979) Chemiluminescence enzyme immunoassay of cortisol using peroxidase as label. *Analytical Biochemistry*, **97**, 248–54.

BARNARD, G., COLLINS, W. P., KOHEN, F. & LINDNER, H. R. (1981) The measurement of urinary estriol-16α-glucuronide by a solid-phase chemiluminescence immunoassay. *The Journal of Steroid Biochemistry*, **14**, 941–8.

BROUGHTON, A. & FRAZIER, M. (1978) A quenching fluoroimmunoassay for the aminoglycoside netilmicin. *Clinical Chemistry*, **24**, 1033.

BURD, J. F., WONG, R. C., FEENEY, J. E., CARRICO, R. J. & BOGUSLASKI, R. C. (1977) Homogeneous reactant-labelled fluorescent immunoassay for therapeutic drugs exemplified by gentamicin determination in human serum. *Clinical Chemistry*, **23**, 1402–8.

BURGETT, M. W., FAIRFIELD, S. J. & MONTHONY, J. F. (1977) A solid phase fluorescent immunoassay for the quantitation of the C_3 component of human complement. *Clinica Chimica Acta*, **78**, 277–84.

CAMPBELL, A. K. & PATEL, A. (1983) A homogeneous immunoassay for cyclic nucleotides based on chemiluminescence energy transfer. *The Biochemical Journal*, **216**, 185–94.

CARRICO, R. J., YEUNG, K. K., SCHROEDER, H. R., BOGUSLASKI, R. C., BUCKLER, R. T. & CHRISTNER, J. E. (1976) Specific protein-binding reactions monitored with ligand ATP conjugates and firefly luciferase. *Analytical Biochemistry*, **76**, 95–110.

CURRY, R. E., HEITZMAN, H., RIEGE, D. H., SWEET, R. V. & SIMONSEN, M. G. (1979) A systems approach to fluorescent immunoassay: General principles and representative applications. *Clinical Chemistry*, **25**, 1591–5.

EKEKE, G. I., EXLEY, D. & ABUKNESHA, R. (1979) Immunofluorimetric assay of estradiol-17β. *The Journal of Steroid Biochemistry*, **11**, 1597–600.

ESSER, A. F. (1980) Principles of electron spin resonance assays and immunological applications. In: *Immunoassays: Clinical Laboratory Techniques for the 1980's*, Nakamura, R. M., Dito, W. R. & Tucker, E. S., III (eds), Alan R. Liss, New York, pp. 213–33.

KOHEN, F., PAZZAGLI, M., KIM, J. B., LINDNER, H. R. & BOGUSLASKI, R. C. (1979) An assay procedure for plasma progesterone based on antibody-enhanced chemiluminescence. *Federation of European Biochemical Societies Letters*, **104**, 201–5.

KOHEN, F., PAZZAGLI, M., KIM, J. B. & LINDNER, H. R. (1980) An immunoassay for plasma cortisol based on chemiluminescence. *Steroids*, **36**, 421–37.

LEUVERING, J. H. W., THAL, P. J. H. M., VAN DER WAART, M. & SCHUURS, A. H. W. (1980) Sol particle immunoassay (SPIA). *The Journal of Immunoassay*, **1**, 77–91.

LINDSTROM, L., MEURLING, L., LÖVGREN, T. (1982) The measurement of serum cortisol by a solid-phase chemiluminescence immunoassay. *The Journal of Steroid Biochemistry*, **16**, 577–80.

OLSSON, T., THORE, A., CARLSSON, H. E., BRUNIUS, G. & ERIKSSON, G. (1979*a*) Quantitation of immunological reactions by luminescence. In: *Proceedings International Symposium of Analytical Applications of Bioluminescence and Chemiluminescence*, 1978, Schram, D. & Stanley, P. E. (eds), State Printing and Publishing Inc, Westlake Village, Calif., pp. 421–30.

OLSSON, T., BRUNUIS, G., CARLSSON, H. E. & THORNE, A. (1979*b*) Luminescence immunoassay (LIA): A solid phase immunoassay monitored by chemiluminescence. *The Journal of Immunological Methods*, **25**, 127–35.

PETTERSSON, K., SIITARI, H., HEMMILA, I., SOINI, E., LÖVGREN, T., HANNINEN, U., TANNER, P. & STENMAN, U.-H. (1983) Time-resolved fluoroimmunoassay of human choriogonadotropin. *Clinical Chemistry*, **29**, 60–4.

SCHROEDER, H. R. & YEAGER, F. M. (1978) Chemiluminescence yields and detection limits of some isoluminol derivatives in various oxidation systems. *Analytical Chemistry*, **50**, 1114–20.

SCHROEDER, H. R., CARRICO, R. J., BOGUSLASKI, R. C. & CHRISTNER, J. E. (1976) Specific binding reactions monitored with ligand–cofactor conjugates and bacterial luciferase. *Analytical Biochemistry*, **72**, 283–92.

SHAW, G. J., WATSON, R. A. A., LANDON, J. & SMITH, D. S. (1977) Estimation of serum gentamicin by quenching fluoroimmunoassay. *The Journal of Clinical Pathology*, **30**, 526–31.

SIMPSON, J. S. A., CAMPBELL, A. K., RYALL, M. E. T. & WOODHEAD, J. S. (1979) A stable chemiluminescent labelled antibody for immunological assays. *Nature*, **279**, 646–7.

SMITH, D. S. (1977) Enhancement fluoroimmunoassay of thyroxine. *Federation of European Biochemistry Societies Letters*, **77**, 25–7.

TENGERDY, R. P. (1967) Quantitative determination of antibody by fluorescence polarisation. *The Journal of Laboratory and Clinical Medicine*, **70**, 707–14.

ULLMAN, E. F., SCHWARZBERG, M. & RUBENSTEIN, K. E. (1976) Fluorescent excitation transfer immunoassays, a general method for determination of antigens. *The Journal of Biological Chemistry*, **251**, 4172–8.

URIOS, P., CITTANOVA, N. & JAYLE, M. F. (1978) Immunoassay of the human chorionic gonadotrophin using fluorescence polarisation. *Federation of European Biochemistry Societies Letters*, **94**, 54–8.

VOLLER, A., BARTLETT, A. & BIDWELL, D. E. (1978) Enzyme immunoassays with special reference to ELISA techniques. *The Journal of Clinical Pathology*, **31**, 507–20.

WANNLUND, J., AZARI, J., LEVINE, L. & DELUCA, M. (1980) A bioluminescent immunoassay for methotrexate at the sub picomole level. *Biochemical & Biophysical Research Communications*, **96**, 440–6.

WATSON, R. A. A., LANDON, J., SHAW, E. J. & SMITH, D. S. (1976) Polarisation fluoroimmunoassay of gentamicin. *Clinica Chimica Acta*, **73**, 51–5.

WEERASEKERA, D. A., KIM, J. B., BARNARD, G. J. & COLLINS, W. P. (1983) The measurement of serum thyroxine by solid-phase chemiluminescence immunoassay. *Annals of Clinical Biochemistry*, **20**, 100–4.

WHITEHEAD, T. P., KRICKA, L. A., CARTER, T. J. N. & THORPE, G. H. G. (1979) Analytical luminescence: Its potential in the clinical laboratory. *Clinical Chemistry*, **25**, 1531–46.

WOODHEAD, J. S., SIMPSON, J. S. A., WEEKS, I., PATEL, A., CAMPBELL, A. K., HART, R., RICHARDSON, A., MCCAPRA, F. (1981) Chemiluminescent labelled antibody techniques. In: *Monoclonal Antibodies and Developments in Immunoassay*, Albertini, A. & Ekins, R. (eds), Elsevier Science Publishers, North Holland, pp. 135–45.

ZUK, R. F., ROWLEY, G. L. & ULLMAN, E. F. (1979) Fluorescence protection immunoassay: A new homogeneous assay technique. *Clinical Chemistry*, **25**, 1554–60.

SESSION II

Application to Macromolecules

5

Species Identification of Meat in Raw, Unheated Meat Products

R. L. S. PATTERSON and S. J. JONES

AFRC Meat Research Institute, Langford, Bristol, UK

INTRODUCTION

In food analysis, it is sometimes necessary to determine whether a sample of meat is actually from the stated species of animal. Whilst improvements in technology and processing have led to more economical transportation and utilisation of deboned carcass meat, they have also facilitated the use of cheaper undeclared meats, because unequivocal identification of species becomes very difficult once meat has been taken off the carcass and the anatomical features have been destroyed. Meat of similar pigmentation, for example beef and horse meat, beef and mutton, or poultry and pig meat, are virtually impossible to distinguish by eye once they have been frozen *en masse* in large blocks, or flaked and incorporated into comminuted meat products. However, apart from deliberate misrepresentation, many meat products in Europe, as well as British-style sausages, burgers and pies, may contain the flesh of more than one species, and the analyst requires rapid methods of species identification.

Methods for the determination of species origin have been available for some time based upon immunological antigen–antibody reactions using various forms of the precipitin test. Precipitating antisera for the different meat species are available commercially and can be used qualitatively in Ouchterlony-type double immunodiffusion tests (Ouchterlony, 1948) or semi-quantitatively in 'rocket' immunoelectrophoresis (Laurell, 1966). One disadvantage of these methods is that they require high amounts of specific antibody preparation in the appropriate test solution or gel to obtain visible precipitin lines. This becomes very expensive in large-scale testing. Gel electrophoresis (Thompson, 1968) and isoelectric focussing (Kaiser *et al.*, 1980; Sinclair & Slattery, 1982) are alternative methods which have had considerable success in identification of the species origin of fresh meats and fish, but which are unsuitable for quantitative analysis of mixtures containing the flesh of more than one species.

Enzyme-linked immunosorbent assay (ELISA) (Engvall & Perlmann, 1971; van Weemen & Schuurs, 1971) has emerged in the last 10 years as a rapid, convenient and relatively cheap method of assaying antigens and antibodies quantitatively in many diagnostic tests in clinical medicine. Recently, we adapted a particular form of ELISA for use in meat species identification (Kang'ethe, Jones & Patterson, 1982). The same technique has also been applied to the detection and estimation of soya protein in food products (Hitchcock *et al.*, 1981; see also Chapter 7), and it is clear that many more applications will be developed in future wherever food components are capable of acting as immunogens.

In clinical analyses samples are usually presented as serum or plasma and in most cases may be added direct to the assay without treatment. In contrast, food is frequently solid, or semi-solid, and discontinuous and some form of extraction is required. Such extracts are likely to have a complex composition and if the extract is used directly the coextracted substances or the solvent *per se* may interfere with antibody and antigen interaction. Additional complications may arise if the key interferent varies in concentration from sample to sample. We have overcome these problems with protein-detecting ELISAs by adsorbing the analyte in the extract onto the walls of the microtitre wells, and then washing the plate to remove solvent and non-adsorbed components. The assay can then be performed using a buffer which optimises antigen–antibody interaction.

This chapter reports the development of a method for identification of the species origin of raw, unheated meats and meat products based on the detection and measurement of the serum albumins present in the sample.

EXPERIMENTAL

Many antisera to animal proteins are now commercially available. Although produced in a host animal (e.g. rabbit, sheep, goat) in response to injection of a single immunogenic substance, for example, a serum albumin, they may nevertheless comprise a mixture of antibodies depending upon the number of antigenic determinants (sites) present on the injected immunogen. Such antisera are termed polyclonal antisera and may require purification by immunoadsorbent chromatography (Kamiyama *et al.*, 1978) before use.

In our method, this procedure consisted of coupling the corresponding antigens such as beef serum albumin (BSA), horse serum albumin (HSA) and sheep serum albumin (SSA) to cyanogen bromide-activated Sepharose 4B and pouring the resulting suspensions into individual chromatography columns. Each column was then equilibrated with 0·15 M phosphate buffered saline (PBS) at pH 7·2. Isolation of the 'monospecific' antibodies from the crude antiserum was achieved by circulating a small volume (<10 ml to avoid overloading) of the polyclonal antiserum, say anti-HSA, through the columns packed with the other two immunoadsorbents (BSA and SSA), continuously for 48 h. Once cross-reacting antibodies to BSA and SSA had been removed, the now diluted anti-HSA serum was circulated through the HSA-immunoadsorbent column for 24 h during which time antibodies to HSA were bound to the column. Any unbound protein remaining in the column was washed out with PBS and the adsorbed antibodies, now monospecific to HSA, recovered by elution from the adsorbent with glycine–HCl buffer (0·1 M, pH 2·5) into 2 M K_2CO_3 to reduce denaturation. The pH value was returned immediately to 7·2 in the pooled antibody containing fractions by addition of solid Tris and then extensive dialysis against PBS. After concentration the solution of monospecific antibodies was stored frozen at −20°C. The cross-reacting antibodies retained by the BSA and SSA immunoadsorbents were removed from the other two columns in the same way and the columns regenerated. Specific antibodies to BSA and SSA were obtained by similar procedures, using the same immunoadsorbents, first by removal of cross-reacting antibodies and then the immobilisation and elution of the specific antibodies.

Extracts of meat or meat mixtures were prepared by homogenisation in saline (0·85% w/v) of the finely minced material, followed by centrifugation at 10 000 rpm for 30 min at 4°C. The supernatant was then filtered through Whatman No. 3 paper to remove floating particles of fat, and stored frozen in 3 ml aliquots or in freeze-dried form.

The extract, now containing the serum albumins residing in the musculature after slaughter, was further diluted in carbonate–bicarbonate coating buffer (0·05 M, pH 9·6), to 1:50 w/v from the meat and dispensed by doubling dilution into adjacent rows of wells on a micro-ELISA plate, and the plate incubated for 3 h at 4°C to allow adsorption of the antigens on to the polystyrene surface. After washing four times with PBST (phosphate buffer-saline containing Tween 20, 0·15%), the species-specific antisera, also diluted in PBST, were added to appropriate test wells and incubated for 2 h at 20–25°C. The plates were then washed again. Since all the antisera had been raised in rabbits, the purified antibodies were therefore rabbit immunoglobulins. The enzyme used to visualise and quantify the assay was horseradish peroxidase which had been conjugated to a second antibody raised against rabbit immunoglobulin in a goat (i.e. an anti-antibody). This provided a common means of detection of all these particular species-specific antibodies. After incubation overnight at 4°C with the antibody–enzyme conjugate, the plates were washed again. The substrate (*o*-phenylenediamine + hydrogen peroxide) was added and allowed to react for 30 min; the reaction was stopped by addition of 12·5% sulphuric acid. The intensity of the yellow colour, measured as absorbance at 492 nm in a micro-ELISA plate reader, was directly proportional to the antigen content of the meat extract.

RESULTS AND DISCUSSION

Immunoadsorbent chromatography was successful in reducing the cross-reactivity of most of the sera; for example, in Ouchterlony tests the crude antiserum to BSA cross-reacted strongly with pure SSA and extracts of sheep, goat and deer meat (venison) prior to chromatography, but not afterwards. However, it should be noted that the antisera, although now with greatly improved specificity, were not necessarily fully monospecific; whilst the anti-horse and the anti-beef serum did not show cross-reactions with the opposing antigens (serum albumins) or with sheep, goat or pig antigens, the 'purified' anti-sheep still cross-reacted strongly with goat, and weakly with beef and deer antigens. Further purification by improved affinity chromatography, or development of a monoclonal antibody, might be successful in reducing the degree of cross-reactivity in the serum. Unfortunately, monoclonal antibodies are not yet available commercially for identification of the common species of meat.

Horsemeat was clearly differentiated from beef, pork and lamb at the optimum dilution of the meat extracts (Fig. 1(a)), and was also detected easily in mixtures with beef at levels above 3% (Fig. 1(b)); however, mixtures containing greater than 80% horsemeat could not be differen-

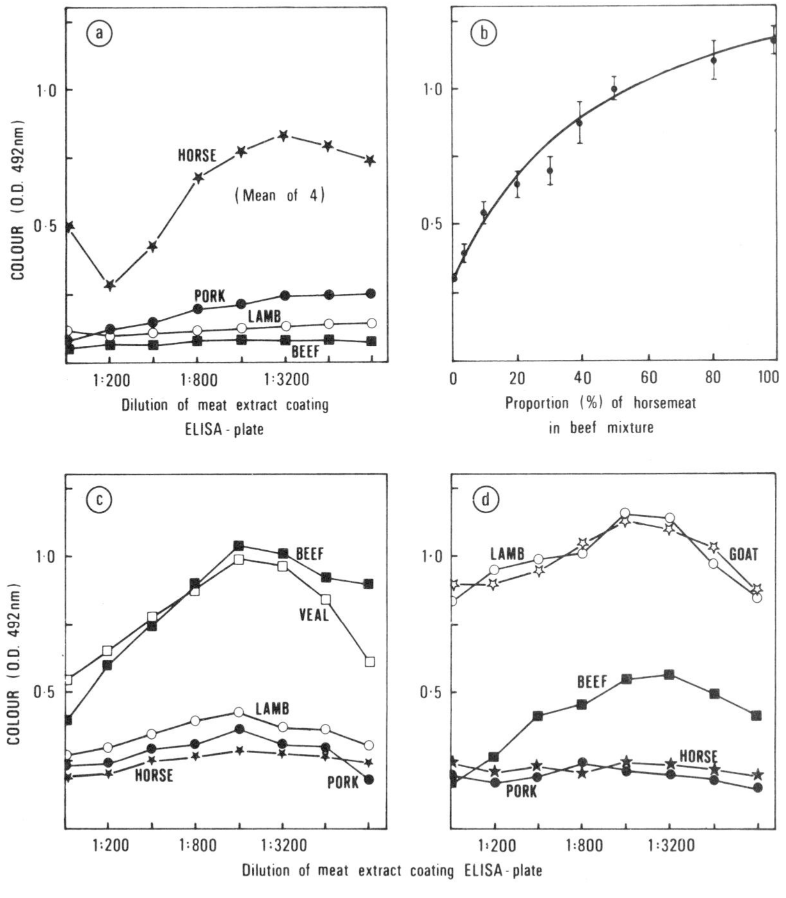

Fig. 1. Identification of meat species by ELISA. (a) Clear differentiation of horsemeat from pork, lamb and beef using an antiserum (adsorbed) to horse serum albumin; (b) increasing colour development as horsemeat content of beef mince increases from 3% using an antiserum to horse serum albumin; (c) differentiation of beef and veal from pork, horse and lamb by an antiserum to bovine serum albumin; (d) cross-reaction of sheep and goat meats (strong) and beef (weak) with an anti-sheep serum. Note that the dilution factor for the meat extract coating the ELISA plate must be optimised to achieve maximum differentiation.

Table 1

Summary of ELISA Responses to Pure Species Albumin in Solution and Meat Extracts of Known Species Origin

Rabbit antiserum to:	*Antigen coating ELISA plate:*						
	BSA Beef	*HSA Horse*	*SSA Sheep*	*Goat*	*PSA Pig*	*Deer*	*Rabbit*
Horse albumin:							
Crude antiserum	+	+ +	+	+	+	+	–
Adsorbed	–	****	–		–		
	–	+ + +	–	–	–	–	–
Bovine albumin:							
Adsorbed	***	–	–		–		
	+ + +	–	+	+	+	+	–
Sheep albumin:							
Adsorbed	–	–	*****		–		
	+	–	+ + + + +	+ + + + +	–	+	+

*, Albumin; +, meat extract.

OD > 0·25 considered positive (* or + for 0·25 unit steps), < 0·25 negative (–).

Results were adjusted to allow for background OD of PBS buffer blank, non-immune serum and conjugate blank.

tiated unambiguously from pure horsemeat using anti-horse serum. Although small differences in absorbance values were found for beef containing less than 3% horsemeat compared with those for pure beef, they were not statistically significant. Greater efficiency of extraction of the serum albumins from the muscle mass would probably improve sensitivity at low concentrations; also introduction of an 'inhibition' stage in the ELISA procedure, which results in an inverse relationship between antigen concentration and intensity of colour developed, would improve sensitivity at low levels. Figure 1(c) shows clear differentiation of beef and veal from pork, horsemeat and lamb with the anti-BSA serum. The cross-reactivity of the anti-SSA serum with goat serum albumin is seen in Fig. 1(d), due to the very close phylogenic relationship of the two species; a degree of cross-reactivity of this antiserum with beef serum albumin is also evident. The presence of rabbit meat in a meat mixture did not interfere with the assay for other species. The extent of the various cross-reactions are summarised in Table 1.

These results demonstrate that use of conventional sera allows species identification of mixed meats; however, such sera may not be fully monospecific even after purification by immunoadsorbent chromatography and some care should be exercised when interpreting the results. It should be noted that the use of antisera raised against serum albumins or most other native proteins will be effective only in the differentiation of raw, unheated meats; a completely new and different set of antisera specific to heat-stable proteins will be required for heat processed and cooked meats.

REFERENCES

ENGVALL, E. & PERLMANN, P. (1971) Enzyme-linked immunosorbent assay (ELISA). Quantitative assay of immunoglobulin G. *Immunochemistry*, **8**, 871–4.

HITCHCOCK, C. H. S., BAILEY, F. J., CRIMES, A. A., DEAN, D. A. G. & DAVIS, P. J. (1981) Determination of soya proteins in food using an enzyme-linked immunosorbent assay procedure. *Journal of the Science of Food and Agriculture*, **32**, 157–65.

KAISER, K.-P., MATHEIS, G., KMITA-DURRMANN, C. & BELITZ H.-D. (1980) Identification of animal species in meat, fish and derived products by means of protein differentiation with electrophoretic methods. (1) Raw meat and raw fish. *Zeitschrift für Lebensmittel Untersuchung und Forschung*, **170**, 334–40.

KAMIYAMA, T., KATSUBE, Y. & IMAIZUMI, K. (1978) Serological identification of animal species of meat using species-specific anti-serum albumin antibodies obtained by immunoadsorbent chromatography. *Japanese Journal of Veterinary Science*, **40**, 663–9.

KANG'ETHE, E. K., JONES, S. J. & PATTERSON, R. L. S. (1982) Identification of the species origin of fresh meat using an enzyme-linked immunosorbent assay procedure. *Meat Science*, **7**, 229–40.

LAURELL, C.-B. (1966) Quantitative estimation of proteins by electrophoresis in agarose gel containing antibodies. *Analytical Biochemistry*, **15**, 45–52.

OUCHTERLONY, O. (1948) *In vitro* method for testing the toxin-producing capacity of diphtheria bacteria. *Acta Pathologica Microbiologica Scandinavia*, **25**, 186–91.

SINCLAIR, A. J. & SLATTERY, W. J. (1982) Identification of meat according to species by isoelectric focusing. *Australian Veterinary Journal*, **58**, 79–81.

THOMPSON, R. R. (1968) An enzymic (esterase) method for identification of animal and fish species. *Journal of the Association of Official Analytical Chemists*, **51**, 746–8.

VAN WEEMEN, B. K. & SCHUURS, A. H. M. W. (1971) Immunoassay using antigen–enzyme conjugates. *FEBS Letters*, **15**, 232–6.

6

Identification of the Species of Origin of Meat in Australia by Radioimmunoassay and Enzyme Immunoassay

L. A. Y. Johnston, P. D. Tracey-Patte, R. D. Pearson,

CSIRO, Division of Tropical Animal Science, Indooroopilly, Queensland, Australia

J. G. R. Hurrell and D. P. Aitken

Commonwealth Serum Laboratories, Parkville, Victoria, Australia

INTRODUCTION

The recent discovery of several instances of substitution of kangaroo and horse meat for beef in 1981, simultaneously in the USA and Australia, led to a Royal Commission into this scandal. While the Royal Commission was in progress, regulations were established requiring species testing of all frozen bulk meat exported from Australia. The samples are obtained from cartons of frozen meat using a twist drill. Several samples are taken from each carton resulting in 25–100 g of meat which resembles minced meat. The most commonly used test for screening meat samples is the agar gel immuno-diffusion test (AGDT). Its routine use has recently been reported by Swart & Wilks (1982) and Kurth & Shaw (1983). The AGDT which is sensitive to approximately 10% w/w contamination and requires at least 6 h for precipitin lines to develop, cannot distinguish between sheep and goat, nor between buffalo and beef meats. Resolution of these species can be accomplished by electrophoretic methods, the most sensitive being agarose isoelectric focusing followed by staining for specific enzymes (King & Kurth, 1982; Sinclair & Slattery, 1982), the detection of contaminants at the 1% level being possible. The main drawback of these techniques is the cost of large volumes of antisera used in the AGDT and gels and substrates in the electrophoretic methods.

Recently the use of radioimmunoassay (RIA) (Johnston *et al.*, 1982) and the potential of enzyme immunoassay (EIA) (Whittaker *et al.*, 1982) for meat identification have been reported in Australia. This paper describes these two techniques and outlines improvements which could be made.

RADIOIMMUNOASSAY

The work reported on RIA was done at the CSIRO Long Pocket Laboratories. Two techniques were used: a direct test in which antigen was reacted with specific antisera labelled with ^{125}I or an indirect test in which antigen was reacted with specific antisera then with protein A labelled with ^{125}I.

Antigen

Samples were taken from frozen meat using a drill and 20 g of the resultant mince was mixed with 30 ml of 0·85% saline and pounded for 1 min in a stomacher. The liquid portion was centrifuged at 2000 g for 20 min and the supernatant used as antigen.

Preparation of Bovine or Buffalo Serum Immunoglobulin (Ig)

Ig was isolated from normal bovine or buffalo serum using the technique described by Mostratos & Beswick (1969). Briefly, a major portion of the albumin and the α- and β-globulins was precipitated from serum using a solution of 0·4% rivanol (ethodin:2-ethoxy-6,9-diaminoacridine lactate). After removing rivanol from the supernatant with activated charcoal (1·2 g/100 ml), the Ig was precipitated by addition of an equal volume of a saturated aqueous solution of ammonium sulphate. The precipitate was redissolved in saline and dialysed for 48 h at 4°C against 0·012 M phosphate-buffered saline (PBS) pH 7·2 in order to remove sulphate ions.

Preparation of Bovine or Buffalo Serum Ig $F(ab^1)_2$

This was prepared by the method of Fey *et al.* (1976). Bovine or buffalo serum Ig was digested with pepsin (2 mg per 100 mg Ig) at 37°C for 24 h. Undigested Ig was removed by precipitation with zinc sulphate to a final concentration of 25 mM. Sulphate ions were removed by dialysing the supernatant containing $F(ab^1)_2$ against 0·012 M PBS pH 7·2.

Preparation of Rabbit Serum Ig

The Ig from immunised rabbits' serum was isolated by precipitation with rivanol and ammonium sulphate as described above.

Preparation of Rabbit Serum against Bovine Colostral Whey Ig

Rabbit antiserum was prepared against bovine colostral whey Ig as described by Lascelles & McDowell (1970). Whey from milk collected from mammary glands which had been infused with killed *Brucella abortus* organisms 3–4 weeks before calving was mixed with *Brucella* cells in antibody excess. The washed sensitised *Brucella* organisms were emulsified with Freund's complete adjuvant (FCA) and injected into rabbits in multiple intramuscular sites. The course of injections was repeated 1 month later and the rabbits bled after a further 2–3 weeks.

Preparation of Rabbit Serum against Bovine Serum Ig

Bovine serum Ig was precipitated with rabbit anti-bovine Ig serum at 37°C for 1 h and allowed to stand at 4°C overnight. The immune complex was washed three times in 0·012 M PBS pH 7·2 emulsified with FCA and used to immunise rabbits as previously described.

Preparation of Rabbit Serum against Buffalo Serum Ig

Buffalo serum Ig (10 mg) was coupled to CNBr-activated Sepharose 4B (1 g) (Pharmacia Fine Chemicals AB, Uppsala, Sweden) according to instructions given by Pharmacia. The sepharose-bound Ig was homogenised with FCA and used to immunise rabbits according to the schedule previously described.

Rabbit Serum Flocculation Ability Test

To compare the ability of various rabbit antisera to precipitate purified bovine serum Ig an 'optimal proportions flocculatión test' was performed (Raffel, 1961). A constant amount of rabbit serum was placed in a row of tubes and a decreasing known amount of antigen added. After mixing, the tubes were incubated at 37°C and the first tube in which flocculation developed was considered the region where antibody and antigen existed in immunologically equivalent proportions. The time taken for flocculation to first develop was recorded.

Preparation of $F(ab^1)_2$ Sepharose 4B

Bovine or buffalo serum $F(ab^1)_2$ (10 mg) was coupled to CNBr-activated Sepharose 4B (2g) (Pharmacia Fine Chemicals AB, Uppsala) according to instructions given by Pharmacia.

Purification of Rabbit Serum against Bovine or Buffalo Ig $F(ab^1)_2$

Rabbit serum against bovine or buffalo Ig was equilibrated with 0·1 M tris buffer pH 8·5 and passed through a bovine or buffalo $F(ab^1)_2$ Sepharose 4B affinity column (1·6 cm × 3·0 cm). The unbound protein was discarded and the bound rabbit anti-bovine or anti-buffalo $F(ab^1)_2$ antibody uncoupled by eluting with 0·2 M glycine–HCl buffer pH 2·3 containing 1·0 M NaCl. The recovered protein was dialysed against 0·012 M phosphate buffer pH 7·2. The protein concentration, as determined by absorbance at 260 nm and 280 nm (Layne, 1957), was adjusted to 5·0 mg/ml.

Rabbit Anti-horse Serum

This serum was obtained from Wellcome Research Laboratories, Beckenham, UK.

Radio-labelling of Proteins

Proteins were labelled by the lactoperoxidase method (Marchalonis, 1969). Briefly, 10 μl of 1 mg/ml protein in 0·05 M phosphate buffer pH 7·4 were coupled with 37 MBq of ^{125}I in a glass tube containing 8·15 μg dipotassium EDTA, 10 μg lactoperoxidase and 250 ng hydrogen peroxide. After 10 min at 23°C the mixture was passed through a column of Sephadex G25M equilibrated with 0·05 M phosphate buffer pH 7·4 and eluted with the same buffer to remove free iodine. The ^{125}I-labelled protein was then aliquoted and stored at −20°C until used. Usual activity was 60–100 000 counts per min/5 ng protein.

Assay

The direct RIA was performed as described by Johnston *et al.* (1982). Dilutions of antigen which were used are indicated in the results.

The indirect RIA was performed by adding various dilutions of meat extract in 0·1 M carbonate buffer pH 9·6 (CB) into each well of a flexible polyvinyl chloride microtitre tray. After 5–6 h at 23°C the wells were washed three times with 0·05 M phosphate-buffered saline pH 7·4 containing 0·1% Tween 20 (PT). Then 200 μl/well of 0·1% gelatine in CB were added. After 1 h at 23°C the wells were washed three times with PT Rabbit serum (200 μl) containing antibodies specific for a meat species at 1:100 dilution in PT + 0·1% gelatine was added and incubated overnight at 4°C. After three washings with PT, 200 μl of ^{125}I-labelled protein A (Pharmacia Fine Chemicals AB, Uppsala) in PT + 0·1% gelatine were added to each well and incubated for 5–6 h at 23°C. Following three washes with PT the wells were cut out and counted in a gamma-counter.

RESULTS

Selection of Rabbit Antiserum for Direct RIA

Not all antisera prepared against bovine Ig were suitable for use in the direct RIA. Table 1 shows the results of testing antisera from 8 different rabbits immunised against bovine IgG. Their sera had been subjected to rivanol then ammonium sulphate precipitation to produce rabbit Ig and after dialysis against 0·05 M PBS pH 7·5 their concentrations of Ig were adjusted to 5 mg/ml. After labelling with ^{125}I, their cpm were adjusted to approximately 250 000 and then they were assayed against bovine IgG at an initial concentration of 10·9 mg/ml. It can be seen that only serum

Table 1
The Relationship Between the Comparative Flocculation Test and Efficiency of the Antiserum in a Direct Radioimmunoassay (RIA)

Rabbit no.	*Comparative flocculation test*		*Counts per min in direct RIA*			
	Bound antigen (μg)	*Time (min)*	*Dilution of bovine immunoglobulin (Ig)* 1:40 000	1:80 000	1:160 000	*Saline*
10	900	0·25	17 003	10 738	6 514	93
29	100	25	1 704	927	680	100
30	100	15	2 865	1 831	1 511	98
51	200	4	2 610	1 348	943	117
54	150	7	1 246	817	632	90
59	100	13	740	460	338	195
60	200	4	2 804	1 582	1 134	176
61	400	1	4 849	2 383	1 715	190

Rabbit 10 was immunised with bovine colostral whey Ig and all the others with bovine serum Ig.
All the sera produced good precipitin lines in double immunodiffusion against bovine Ig.

from rabbit 10 was suitable for direct RIA. Two of these sera (54 and 61) were applied to an affinity column containing the $F(ab^1)_2$ portion of bovine Ig, eluted with glycine–HCl, dialysed against PBS and then labelled with ^{125}I. Results of RIA tests of these sera before and after affinity chromatography are given in Table 2 and show that sera with titres lower than that of rabbit 10 were suitable for use in RIA after purification by affinity chromatography.

Table 2
The Effect of Affinity Purification of Antibodies on the Sensitivity of the Direct Radioimmunoassay (RIA)

Rabbit no.		*Counts per min in direct RIA* — *Dilution of bovine immunoglobulin* *1:40 000*	*1:80 000*	*1:160 000*
54	Before chromatography	1 248	470	143
54	After chromatography	13 457	2 411	424
61	Before chromatography	7 078	2 060	390
61	After chromatography	36 444	7 242	1 184

The rabbit numbers are the same as in Table 1.

Direct RIA

Figure 1 shows the results of an experiment where dilutions of meat extract were reacted with an Ig fraction of rabbit serum containing antibody to the $F(ab^1)_2$ portion of bovine Ig (*Bos taurus*) and labelled with ^{125}I (Johnston *et al.*, 1982). At very high dilutions, cattle meat can be differentiated from sheep, donkey, pig, horse and kangaroo meat. This test could be used as a screening test and any samples which gave low counts could be examined by the indirect RIA or electrophoretic analysis to determine which contaminating meats were present.

In practice, a dilution of meat extract is made in microtitre wells and then the test is performed. Johnston *et al.* (1982) reported that this was successful in identifying beef samples and differentiating them from 100% horse and kangaroo meat. If contaminating meats are to be screened, then for each species the antisera against the $F(ab^1)_2$ part of Ig labelled with ^{125}I are required for the direct RIA. This has been done for buffalo meat with the results shown in Table 3.

Antisera for other species for which tests are required are currently being prepared but this approach shows most promise for being the RIA test of choice. Given that the processing of the meat samples is the same for all tests, then the major expense incurred will be in cost of antisera and using the direct RIA, 10 μl of a specific antiserum for the $F(ab^1)_2$ portion of Ig will enable 30 000 tests to be done.

Indirect RIA

This test has been done using commercially available antisera and the basis of the antigenic determinants used is unknown. The test accurately

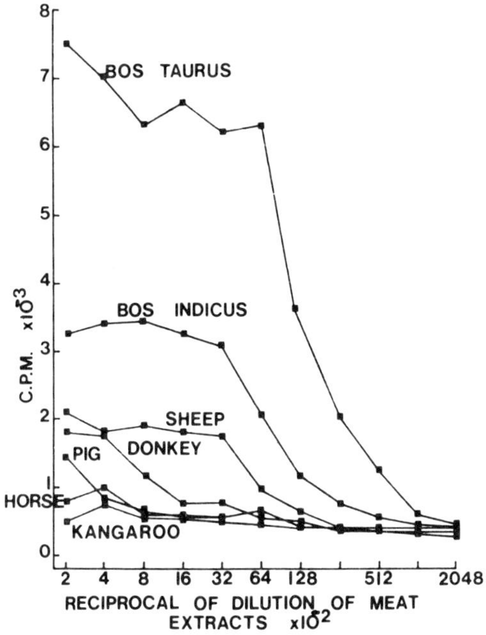

Fig. 1. Mean counts per min (cpm) from a muscle extract of *Bos taurus*, *Bos indicus*, horse, donkey, kangaroo, pig and sheep plotted against the dilution of the meat extracts in a direct RIA using rabbit serum against the $F(ab^1)_2$ portion of bovine Ig (*Bos taurus*). Reproduced with permission from Johnston *et al.* (1982).

Table 3

Results of Testing Bovine and Buffalo Meat Using Direct Radioimmunoassay (RIA) with Two Rabbit Sera, one Containing Antibodies to Bovine Meat and the other Antibodies to Buffalo Meat

	Bovine meat (cpm in direct RIA)	*Buffalo meat (cpm in direct RIA)*
Bovine antiserum	11 575	2 939
Buffalo antiserum	2 304	4 094

detects contamination to at least the 5% level of species other than beef in a binary mixture. The meat mixture at a dilution of 1:800 in microtitre wells was reacted with rabbit anti-horse serum and then with protein A labelled with ^{125}I. A typical example is shown in Table 4.

Table 4
Results of Testing Varying Mixtures of Horse and Bovine Meat Using Indirect Radioimmunoassay with Rabbit Serum Containing Antibodies to Horse Meat

Horse in beef (%)	0	1	5	10	25	50	100
Counts per min	2 390	2 331	4 347	6 803	7 559	10 286	11 071

Other species have been tested and these results will be reported when the work is finalised. This test is similar to EIA and 2 µl of an antiserum containing 4–5 mg/ml of protein are required for each test.

ENZYME IMMUNOASSAY

The work reported on EIA was done at Commonwealth Serum Laboratories, Parkville, Australia.

Assay

The formats used for RIA procedures (competitive, indirect, capture or sandwich) are also applicable for EIA providing that methods used for separating bound and free labelled reagent are compatible with maintenance of enzyme activity. The solid-phase indirect and capture EIA methods have been described for the identification of meat species (Kang'ethe *et al.*, 1982; Patterson *et al.*, 1983) with sensitivity generally quoted as between 1% and 10% w/w in known artificial mixtures. The capture* procedure is the more consistent and sensitive of the two procedures and gives better resolution between closely related species (Kang'ethe *et al.*, 1982; Patterson *et al.*, 1983). The better performance of the capture procedure for meat species testing is probably due to the consistency of preparation of the solid phase capture surface and the two site species discrimination required in this format. The indirect assay uses meat extracts, which are variable in molecular and species composition, to coat the solid phase, most usually the wells of a polyvinylchloride (PVC) disposable microtitre 96 well tray. The species-specific antibody brought into the wells recognises any of that species antigen adsorbed to the PVC and the enzyme-labelled anti-species IgG antibody detects species-specific antibody complexed to the PVC-bound antigen. The capture assay employs a species-specific antibody, adsorbed to the PVC

*This approach is also known as the 'sandwich' or 'two-site' assay.

wells, to bind meat antigen and enzyme-labelled species-specific antibody to indicate the presence of bound antigen. The meat species identification test described here employs the capture assay configuration (Aitken *et al.*, 1983).

Enzyme Label

The enzymes most frequently used to label the tracer reagent in EIA procedures are horseradish peroxidase (HRP) and alkaline phosphatase (AP). Kang'ethe *et al.* (1982) developed an indirect EIA test for meat species identification using HRP-labelled antibodies whilst Patterson *et al.* (1983) have preferred AP-labelled antibodies for their capture assay procedure. Unfortunately, for a rapid, field type assay kit these enzymes have several disadvantages. Both use substrates which are unstable in aqueous solution, hence substrate solutions must be freshly prepared just prior to use; they have diffuse endpoints in titration experiments (particularly AP) and neither have substrates which give an easily recognised colour change. Recently, we introduced the enzyme urease and a substrate solution consisting of urea and the pH indicator bromocresol purple for EIA use (Chandler *et al.*, 1982) which overcomes many of the disadvantages seen with HRP and AP. Urease gives a highly visual colour change (yellow to purple) with sharp endpoint when used with the stable urea-indicator substrate solution making it particularly useful for mass screening and field-type EIA procedures. The system we have developed for meat species identification employs urease-labelled antibodies and the urea-pH indicator substrate solution.

Antibody Specificity

The antisera used in the AGDT procedure for meat testing are often raised against mixtures of antigens such as crude meat extracts or whole serum. These are usually very good precipitating sera; however, they cause difficulties when used in the sensitive and quantitative enzyme immunoassays. The mixed nature of the immunogen causes inconsistency of response in rabbits, creating problems of test standardisation and making difficult the removal of antibodies cross-reacting with a number of species. A homogeneous immunogen provides a more consistent immunological response in the immunised animal and the level of cross species reaction is reduced and more easily removed.

The two most obvious immunogens for raising antibodies for meat species testing are serum albumin and IgG. Kang'ethe *et al.* (1982) selected albumin as the immunogen and produced effectively monospecific antisera by immunoadsorption on sepharose albumin columns.

These affinity purified antisera allowed most species to be distinguished, however, some low level of cross-reaction remained.

We chose to use IgG as the immunogen. This selection was based mainly on our extensive experience in the purification of immunoglobulins from a large number of species and the raising of specific antisera. The phylogenetic differences (both amino acid sequence and immunological) in these molecules are well documented (Kubo *et al.*, 1973). Antisera based on the immunoglobulins also showed extensive cross-reaction with a number of species, and this cross-reaction was not always reciprocal. This cross-reaction necessitated the use of sequential affinity purification steps. The antisera produced by affinity purification allowed the development of assays for beef, goat, horse, kangaroo and buffalo meat at 1% w/w level of contamination within 2 h with no obvious cross-reaction. Prolonged development of the assay with substrate did show very low levels of cross-reaction. No satisfactory specific antisera could be raised and purified for sheep and pig meat assays. The level of cross-reaction with anti-sheep IgG was so great (particularly with beef) that affinity purification removed even the antibodies against sheep IgG. Kang'ethe *et al.* (1982) also experienced difficulty in removing beef, goat and deer cross-reacting antibodies from antisera raised against sheep albumin.

The levels of cross-reaction observed with sheep anti-beef and rabbit anti-beef antibodies are shown in Table 5. The raising of anti-beef antibodies in sheep completely abrogated cross-reaction with sheep and led to a very sensitive assay. Two passes down a buffalo IgG affinity column removed 65 mg of antibody from 100 mg of unpurified anti-beef antibody. Passage of the remaining 35 mg of antibody down horse and pig IgG affinity columns removed 6 mg and 2 mg respectively, giving 27 mg of monospecific sheep anti-beef IgG. This antibody was used to develop a highly specific and sensitive assay for beef meat in the presence of meat from other species. The rabbit anti-sheep antibody could not be purified to give a monospecific antibody. Passage down goat, beef and horse IgG affinity columns removed anti-sheep activity as well as the other species giving rise to an insensitive, non-specific assay. A similar situation was observed with rabbit anti-pig IgG which reacted with beef, sheep, horse, buffalo and kangaroo meat extracts. Affinity purification failed to remove the cross-reactions completely.

The difficulty of cross-reaction seen with both anti-sheep IgG and anti-pig IgG may be overcome by the use of monoclonal antibodies produced by murine somatic cell hybrids (hybridomas) and work on producing specific antibodies for various Ig-species is currently underway.

Table 5
Cross Reaction of Sheep Anti-beef Immunoglobulin G (IgG) and Rabbit Anti-beef IgG with Meat Extracts in a Capture EIA Before Affinity Purification[a]

Meat extract	*Concentration (% w/w) detected 15 min after addition of substrate solution*	
	Sheep anti-beef IgG	*Rabbit anti-beef IgG*
Beef	0·8	0·4
Pig	75·0	NT
Horse	50·0	50·0
Buffalo	1·6	NT
Sheep	ND	0·1
Goat	ND	0·1
Kangaroo	ND	ND
Camel	ND	NT

ND: Not detected.
NT: Not tested.
[a]Isolated sheep or rabbit IgG containing antibodies to bovine IgG (before affinity purification) was coated at 50 μg/ml onto wells of a polyvinylchloride 96 well plate. Meat extract was introduced into the wells in varying dilutions in PBS-0·5% Tween 20 starting at 100%. Urease-labelled unpurified sheep or rabbit IgG containing antibodies to bovine IgG was the tracer reagent.

Assay Performance

The assay we have developed as a commercial kit is designed to detect contamination at 1% w/w within 2 h total test time. The kit comprises a 96 well flexible PVC plate coated with affinity purified antibody. A meat extract, prepared by vortex mixing chopped meat with PBS-0·5% v/v Tween 20 (1 g/1·5 ml), is incubated for 30 min at 37°C in the antibody-coated wells. After washing, the wells are incubated with urease-labelled anti-species IgG antibody for 30 min at 37°C. Washing is followed by the addition of urease substrate solution and observation for 15 min at room temperature. The presence of meat of the species under examination is indicated by the vivid yellow to purple colour change of the substrate solution. Standards of known meat composition are assayed simultaneously to provide semi-quantitative information.

A comparison of the performance of the urease EIA for meat species identification with other methods is shown in Table 6. All but two of the samples tested were purchased as 'ground beef' in various capital cities of Australia.

Table 6
Comparisons of Results Obtained with Fifteen Meat Samples using Five Different Procedures to Identify Meat Species

Sample number	*Enzyme immuno-assay (EIA)*	*Agar gel immuno-diffusion test (AGDT)*	*Lactate dehydro-genase (LDH)*[b,f]	*Esterase*[b,e]	*Iso-electric focussing IEF*[c]
1	B,S,P	B,S,P	P	S	B,S,P
2	B,S,P	B,S	–	S	B,S
3[a]	B	B[d]	NT	–	B
4	B,BF	B	NT	–	B,BF
5	B,BF,P	B	NT	–	B,BF
6	B,S,P	B,S,P	P	S	B,S,P
7[a]	B,S,BF	B,S[d]	–	S	B,S,BF
8	B,S,P	B,S,P	P	S	B,S,P
9	B,S	B,S,P	–	S	B,S
10	B,S	B,S	NT	S	B,S
11	B,S	B,S	–	S	B,S
12	B,S	B,S	NT	S	B,S
13	B	B	NT	–	B
14	B,S,P	B,S,P	P	S	B,S
15	B,S,P	B,S,P	P	S	B,S

B: beef; S: sheep; P: pig; BF: buffalo.
NT: sample not tested; –: indicates a negative result.
[a]These samples were artificial mixtures set up as control samples.
[b]Gel electrophoresis followed by specific isoenzyme stain.
[c]Isoelectric focussing followed by total protein stain.
[d]AGDT cannot distinguish between beef and buffalo meat.
[e]Esterase activity allows discrimination between sheep and goat meat.
[f]LDH activity confirms the presence of pig meat.

The LDH (lactate dehydrogenase) and esterase (isoenzyme) tests were performed following electrophoresis of the specimens on cellulose acetate (Slattery & Sinclair, 1983). The LDH test was used to confirm the presence of pig meat in samples positive for pig by the AGDT and the esterase test was used to discriminate between sheep and goat meat, a distinction not possible with AGDT. The isoelectric focussing (IEF) test was useful for discrimination between beef and buffalo meats. This was particularly important in three samples which did contain both beef and buffalo meats.

The urease EIA results shown in Table 6 agree in most cases with the other four test results with the exception of sample numbers 2, 5 and 9. The EIA result for sample number 2 indicates the presence of pig meat, however AGDT, LDH and IEF tests were all negative for pig. Similarly, the EIA result for sample number 5 indicates pig meat, whereas the AGDT and IEF tests were negative for pig. These results could be due to the higher sensitivity of the EIA procedure over that of the other tests. However, given the difficulty we have experienced in producing a monospecific, polyclonal anti-pig IgG antibody, it is also possible that the results are due to remaining cross-reactivity. In addition the AGDT indicates the presence of pig meat in sample number 9, whilst the LDH, esterase and EIA procedures all show a negative test for pig meat. These ambiguities regarding the detection of pig meat, as a contaminant, should be resolved by the use of highly specific monoclonal antibodies against pig IgG. A feature of the comparisons shown in Table 6 was the detection of buffalo meat as a contaminant of beef. This distinction is not possible with the AGDT usually employed for rapid, mass screening.

FUTURE DIRECTIONS FOR IMPROVING RIA AND EIA

The direct RIA using rabbit antisera against $F(ab^1)_2$ immunoglobulin is probably worth developing. If specific antisera are generated against all the species which may contaminate beef then this test will use such small quantities of antisera that the cost per test will be small even though expensive equipment and trained staff are necessary. The indirect RIA can be improved by the generation of specific antisera but it will only be as good as EIA.

Undoubtedly, the main thrust of future development of the EIA procedure for meat testing will be in the area of improving the antibody specificity and avidity. The application of hybridoma technology to the production of monoclonal antibodies for meat species testing will improve the antibody quality (specificity and possibly, by adequate selection and cloning, avidity) whilst removing the requirement for tedious and expensive affinity purification and extensive testing. Once the hybridoma cell line secreting antibody with the appropriate characteristics has been selected, we will have an unlimited supply of antibody of invariant and optimal quality. In addition to the constancy of antibody quality, the introduction of hybridoma-derived monoclonal antibodies with no inter-species cross-reactions will open up the possibility of simultaneous multi-species testing to determine if meat substitution or contamination has occurred without identifying the exact offending species. This test could be carried out on-site at inspection points to allow rapid shipment or sale of uncontaminated meat. Meat found to be positive by the on-site species contamination test could be further tested in the laboratory by species-specific EIA or other non-immunological tests to identify the contaminant species. This approach to testing should speed up the shipment of meat and decrease the number of samples requiring in-depth laboratory analysis.

A further refinement of the multispecies contamination testing may be in the use of capillary tubes in a test similar to that used for the identification of snake venoms in clinical specimens (Chandler & Hurrell, 1982). This test could use capillaries coated with a mixture of monoclonal antibodies against up to four species per capillary (as indicated by unpublished data) for the capture component of the assay and a mixture of urease-labelled species-specific antibodies as the tracer. Preliminary investigation has suggested that this approach is feasible and would be particularly useful for municipal health inspectors, export meat inspection officers and manufacturers of processed meat products as a 'front-line' raw material quality check.

ACKNOWLEDGEMENTS

We wish to thank Frank Shaw of CSIRO Meat Research Laboratories for providing meat samples and Alan Donaldson of CSIRO Long Pocket Laboratories for technical assistance. Our colleagues at the Commonwealth Serum Laboratories: John Cox, Howard Chandler, Andrew MacGregor and Robert Premier, are thanked for their collaboration and advice in the execution of the EIA development programme. Alan Scammell of the Victorian Regional Laboratory of the Australian Government Analytical Laboratories is thanked for his collaboration in performing the AGDT, LDH, esterase and IEF tests shown in Table 6.

REFERENCES

AITKEN, D. P., CHANDLER, H. M., PREMIER, R. & HURRELL, J. G. R. (1983) Meat species identification using urease-antibody. In: *Horizon 90: How to Survive the 80s. 16th Annual Convention Australian Institute of Food Science and Technology*, p. 17.

CHANDLER, H. M. & HURRELL, J. G. R. (1982) A new enzyme immunoassay system suitable for field use and its application in a snake venom detection kit. *Clinica Chimica Acta*, **121**, 225–30.

CHANDLER, H. M., COX, J. C., HEALEY, K., MACGREGOR, A., PREMIER, R. R. & HURRELL, J. G. R. (1982) An investigation of the use of urease-antibody conjugates in enzyme immunoassays. *Journal of Immunological Methods*, **53**, 187–94.

FEY, H., PFISTER, H., MESSERLI, J., STURZENEGGER, N. & GROLIMUND, F. (1976) Methods of isolation, purification and quantitation of bovine immunoglobulins. *Zentralblatt für Veterinische Medizin B*, **23**, 269–300.

JOHNSTON, L. A. Y., TRACEY-PATTE, P. D., DONALDSON, R. A. & PARKINSON, B. (1982) A screening test to differentiate cattle meat from horse, donkey, kangaroo, pig and sheep meats. *Australian Veterinary Journal*, **59**, 59.

KANG'ETHE, E. K., JONES, S. J. & PATTERSON, R. L. S. (1982) Identification of the species origin of fresh meat using an enzyme-linked immunosorbent assay procedure. *Meat Science*, **7**, 229–40.

KING, N. L. & KURTH, L. (1982) Analysis of raw beef samples for adulterant meat species by enzyme-staining of isoelectric focusing gels. *Journal of Food Science*, **47**, 1608–12.

KUBO, R. T., ZIMMERMAN, B. & GREY, H. M. (1973) Phylogeny of immunoglobulins. In: *The Antigens*, Vol. 1, Sela, M. (ed.), Academic Press, New York, pp. 417–77.

KURTH, L. & SHAW, F. D. (1983) Identification of the species of origin of meat by electrophoretic and immunological methods. *Food Technology, Australia*, **35**, 328–31.

LAYNE, E. (1957) Spectrophotometric and turbidimetric methods for measuring proteins. *Methods in Enzymology*, **3**, 447–54.

LASCELLES, A. K. & MCDOWELL, G. H. (1970) Secretion of IgA in the sheep following local antigenic stimulation. *Immunology*, **19**, 613–20.

MARCHALONIS, J. J. (1969) An enzymatic method for the trace iodination of immunoglobulins and other proteins. *Biochemistry Journal*, **113**, 299–305

MOSTRATOS, A. & BESWICK, T. S. L. (1969). Comparison of some simple methods of preparing gamma-globulin and antiglobulin sera for use in the indirect immunofluorescence technique. *Journal of Pathology*, **98**, 17–24.

PATTERSON, M. R., SPENCER, T. L. & WHITTAKER, R. G. (1983) Enzyme linked immunosorbent assays of speciating meat. In: *Horizon 90: How to Survive the 80s. 16th Annual Convention Australian Institute of Food Science and Technology*, p. 16.

RAFFEL, S. (1961) *Immunity*, 2nd edn, Appleton-Century-Crofts, New York, p. 155.

SLATTERY, W. J. & SINCLAIR, A. J. (1983) Electrophoretic methods of species identification of meat. In: *Horizon 90: How to Survive the 80s. 16th Annual Convention Australian Institute of Food Science and Technology*, p. 15.

SINCLAIR, A. J. & SLATTERY, W. J. (1982) Identification of meat according to species by isoelectric focusing. *Australian Veterinary Journal*, **58**, 79–80.

SWART, K. S. & WILKS, C. R. (1982) An immunodiffusion method for the identification of the species of origin of meat samples. *Australian Veterinary Journal*, **59**, 21–2.

WHITTAKER, R. G., SPENCER, T. L. & COPLAND, J. W. (1982) Enzyme-linked immunosorbent assay for meat species testing. *Australian Veterinary Journal*, **59**, 125.

7

The Determination of Soya Protein in Meat Products*

C. H. S. Hitchcock

Unilever Research, Colworth Laboratory, Sharnbrook, Bedford, UK

The application of immunoassays to the determination of food proteins is hampered by the composite nature of the protein ingredient (e.g. soya) and the biological variability of its composition; by the processing involved in the manufacture of the ingredient (e.g. soya flour, concentrate, isolate or extrudate) and the final product (e.g. raw, heat-set, canned, dried) (Olsman & Hitchcock, 1980). The sample may be initially solubilised using protein denaturants (e.g. urea, detergent) and 're-natured' to an antigenic form by dialysis or dilution. Polyvalent antisera against renatured, heat-denatured and/or native protein mixtures may be made the basis of the immunoanalysis; particularly convenient is an enzyme-linked immunosorbent assay (ELISA) in the inhibition mode which requires only commercially available reagents — antiserum, labelled anti-globulin and an appropriate standard soya ingredient (Hitchcock *et al.*, 1981). The response to different commercial soya ingredients is somewhat variable when compared to a single arbitrary standard; indeed,

*Further information published in: (1) A. A. Crimes, C. H. S. Hitchcock & R. Wood (1984) Determination of soya protein in meat products by an enzyme-linked immunosorbent assay procedure: Collaborative study. *Journal of the Association of Public Analysts*, **22**, 59–78; (2) N. M. Griffiths, M. J. Billington, A. A. Crimes & C. H. S. Hitchcock (1984) Enzyme-linked immunosorbent assay (ELISA) of soya protein in meat products using commercially available reagents. *Journal of the Science of Food and Agriculture*, **35**, 1255–60; (3) W. J. Olsman, S. Dobbelaere and C. H. S. Hitchcock The performance of an SDS-PAGE and an ELISA method for the quantitative analysis of soya protein in meat products: An international collaborative study. *Journal of the Science of Food and Agriculture* (in press).

using current antisera the assay is more sensitive to isolated 7S-component of soya protein than to the 11S-protein or the whey protein. Nevertheless, the observed levels of soya protein in raw, heat-set or dried model meat products were reasonably accurate when a single appropriate standard was used. When the samples (but not the standard) were canned at 121°C for 30 min, allowance had to be made for a consistent decrease in response of about 50%. Interference from meat, milk, egg, wheat and field bean proteins is negligible. It is concluded that immunoassays represent a powerful cost-effective method of potential application to food proteins, and indeed to any antigenic analyte. Advantages include reliability, specificity, sensitivity and speed, especially if commercially available in convenient kit form.

REFERENCES

Hitchcock, C. H. S., Bailey, F. J., Crimes, A. A., Dean, D. A. G. & Davies, P. J. (1981) Determination of soya proteins in food using an enzyme-linked immunosorbent assay (ELISA) procedure. *Journal of the Science of Food and Agriculture*, **32** (2), 157–65.

Olsman, W. J. & Hitchcock, C. (1980) Detection and determination of vegetable proteins in meat products. In: *Developments in Food Analysis Techniques*, King, R. D. (ed.), Vol. 2, Applied Science Publishers Ltd, London, pp. 225–60.

8

The Results of a Collaborative Trial to Determine Soya Protein in Meat Products by an ELISA Procedure*

R. Wood

Ministry of Agriculture, Fisheries and Food, London, UK

The results of a collaborative study carried out in 22 UK laboratories using an enzyme-linked immunosorbent assay (ELISA) procedure to determine soya protein in meat products are reported. The method tested is based on work previously reported (Hitchcock *et al.*, 1981). Most of the participating laboratories were not familiar with the method before receiving the trial samples. These were sausage-type products which had been prepared with soya flour or textured soya concentrate at levels of incorporation of between 0 and 7% soya protein.

The results of the trial were statistically analysed by procedures outlined by the British Standards Institution (1979).

The repeatability values obtained were of the order of 30% of the determined value of soya protein for the soya flour samples and 60% for the textured soya concentrate samples. The reproducibility values were of the order of 70% of the determined value of soya protein for both types of soya samples.

The results obtained in the ELISA trial are compared with those obtained in a previous trial to estimate soya material by a stereological procedure (Flint & Meech, 1978). It was found that the results from the trial using the ELISA method are significantly better than those obtained in the previous trial using the stereological procedure which can only be recommended as a screening procedure.

*Full paper published in Crimes, Hitchcock & Wood (1984) – for full reference see footnote, p. 111, ref. (1).

REFERENCES

BRITISH STANDARDS INSTITUTION (1979) Precision of Test Methods, BS 5497, Part 1, London.

FLINT, F. O. & MEECH, M. V. (1978) Quantitative determination of texturised soya protein by a stereological technique. *Analyst*, **103**, 252–8.

HITCHCOCK, C. H. S., BAILEY, F. J., CRIMES, A. A., DEAN, D. A. G. & DAVIES, P. J. (1981) Determination of soya proteins in food using an enzyme-linked immunosorbent assay (ELISA) procedure. *Journal of Science of Food and Agriculture*, **32** (2), 157–65.

9

Determination of Milk Protein Denaturation by an Enzyme-linked Immunosorbent Assay

L. M. J. HEPPELL*

National Institute for Research in Dairying, Shinfield, Reading, UK

INTRODUCTION

A small proportion of babies fed with cow's milk based infant formulas develop allergy to cow's milk proteins. Recently, McLaughlan *et al.* (1981) showed that the antigenicity of cow's milk proteins could be reduced by heat treatment, and that commercial baby milk formulas that had received severe heat treatment during manufacture were less effective at sensitising guinea pigs by mouth than mildly heated preparations.

Work from this laboratory (Kilshaw *et al.*, 1982) established that severe heat treatment of skimmed milk had little effect on its ability to sensitise guinea pigs for systemic anaphylaxis when given orally, but similar heat treatment applied to whey completely abolished its sensitising capacity. We have therefore been investigating the possibility of basing a hypoallergenic infant milk formula on heat denatured whey.

The assessment of antigenicity by feeding experiments in guinea pigs was unsatisfactory for testing large numbers of samples and we therefore developed an assay *in vitro*. Detection of residual antigenic milk proteins after graded heat treatments by a competitive inhibition enzyme-linked immunosorbent assay (ELISA) has been reported in our previous papers (Kilshaw *et al.*, 1982; Heppell *et al.*, 1984). This paper describes the method in detail, and examines the extent to which a test *in vitro* can be used to predict the antigenicity of a protein *in vivo*. Problems encountered when measuring the antigenicity of heat denatured proteins will be discussed.

*Present address (from 1 April 1985): Animal and Grassland Research Institute, Shinfield, Reading, UK.

MATERIALS AND METHODS

Competitive Inhibition ELISA for Quantitation of Milk Proteins

Residual undenatured β-lactoglobulin (βLG), α-lactalbumin, bovine serum albumin, bovine immunoglobulin G_1 and α_{s_1}-casein in the heated samples of milk and whey were measured with specific antisera raised against purified milk proteins by a competitive inhibition ELISA based on the methods and reagents described by Voller *et al.* (1976). Their specificity was checked by immunodiffusion. Dilutions of test samples were mixed with a standard dilution of specific antiserum and the antibodies remaining unbound after a period of incubation were measured by ELISA (Fig. 1).

Serial 4-fold dilutions of the test sample were mixed with equal volumes of a standard dilution of rabbit antiserum to the test protein and incubated overnight at 4°C. Microtitre plates (Dynatech M29) were coated by overnight incubation (pH 9·6, 4°C) with purified milk protein (1 μg/ml). After extensive washing, 200 μl of the sample–antibody mixture were added to each well and the plates incubated at room temperature for 6 h. They were then washed and incubated overnight at 4°C with 200 μl/well of alkaline phosphatase goat anti-rabbit IgG conjugate (Miles Laboratories) at 1/6000. After a final wash, 200 μl of disodium nitrophenyl phosphate solution (0·5 mg/ml) were added to all wells and the optical density (OD) at 405 nm measured after precisely 30 min at room temperature.

For each milk protein, ODs (duplicate assays) were plotted against $\log_2$ of sample dilution. Test samples were assayed in parallel with unheated skimmed milk and purified milk protein at a known concentration, and the number of doubling dilutions separating the midpoints of the assay curves was determined. Results were expressed either as absolute quantities of undenatured protein calculated from the purified protein standard, or as fractions, to the nearest doubling dilution. of the levels in unheated milk.

Preparation of skimmed milk and whey, and the method of heat treatment, have been described previously (Kilshaw *et al.*, 1982). Antigenicity was assessed *in vivo* by the ability of the heated sample to sensitise guinea pigs both for systemic anaphylaxis and serum antibody production when given orally (Kilshaw *et al.*, 1982; Heppell *et al.*, 1984). Briefly, guinea pigs were given heated or unheated milk or whey to drink for 2 weeks. The milk was then replaced by water and 6 days later blood samples were taken. The animals were tested for systemic anaphylaxis on

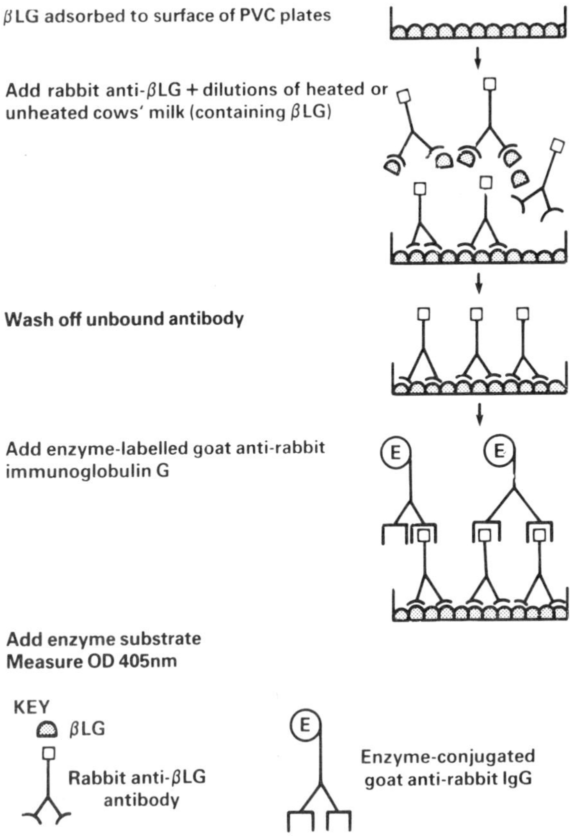

Fig. 1. Competitive inhibition ELISA for β-lactoglobulin (βLG).

the following day by intravenous injection of 0·5 ml of the preparation used for feeding. Serum antibodies to individual proteins were determined by an ELISA method (Heppell & Kilshaw, 1982).

RESULTS AND DISCUSSION

A variety of ELISA methods are available for measuring antigen. We have used a competitive inhibition method designed to utilise a single commercial enzyme-labelled antibody preparation in assays for all milk proteins, rather than separate labelled reagents for each.

Sensitivity

We found that the assay was capable of detecting levels of purified βLG down to 10 ng/ml, as measured from the mid-point of the assay curve.

Precision

Variation in results can occur between duplicates on the same plate, duplicate plates used on the same day, and between results determined on different days. Variation between duplicates on the same plate was rarely more than 10%. To investigate the between-plate and between-day variations, samples of βLG were prepared at concentrations of 5000, 1000, 5 and 1 μg/ml. The amount of βLG in these samples was then assessed by the competitive inhibition ELISA on duplicate plates on two separate days, using the 5000 μg/ml sample as the standard. Results are shown in Table 1. The variation between results obtained on different days was similar to the variation between duplicate plates and was generally less than 20%. In our assays, the error introduced by repeated 4-fold dilutions was small, since the concentrations of βLG determined by ELISA were very similar to the expected values of 1000, 5 and 1 μg/ml.

Table 1
Precision of Competitive Inhibition ELISA for βLG

Concentration of βLG (μg/ml) in prepared samples	*Concentration of βLG by ELISA (μg/ml)* Day 1		Day 2	
	Plate 1	*Plate 2*	*Plate 1*	*Plate 2*
1 000	1 020	1 090	980	900
5	5·2	5·8	4·3	5·2
1	1·2	1·1	1·3	1·0

Assay curves for heated and unheated milk proteins differed in shape (Fig. 2). This could affect the reliability of test results calculated from standard curves, and may be caused by a reduction in the affinity for antibody in partially denatured or heat aggregated molecules. This is a potential problem with other immunological assays of heat-treated proteins which may differ from the unheated protein used as the standard. Nevertheless, the ELISA was capable of discriminating between samples of whey receiving only small increments in the level of heat treatment (Fig. 3).

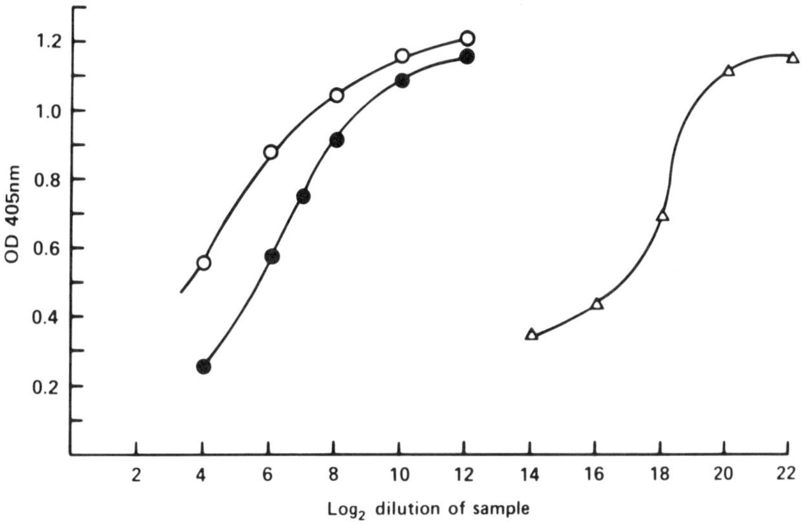

Fig. 2. Assay of βLG in test samples of unheated skimmed milk, △; and whey heated at 100°C for 30 min, ●; or 115°C for 30 min, ○.

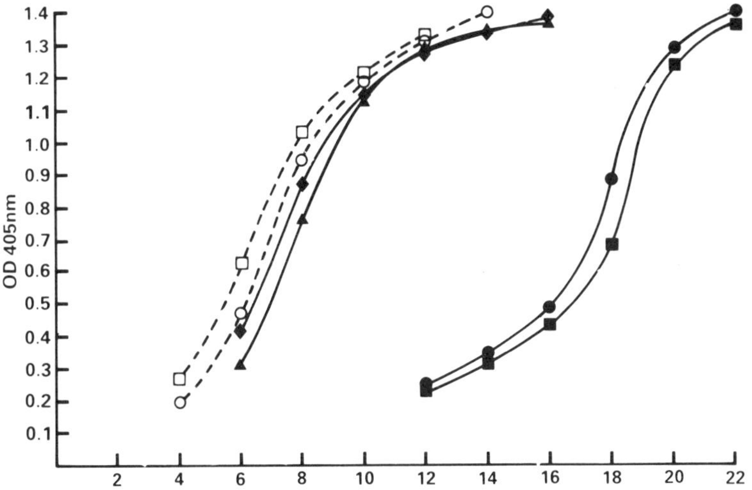

Fig. 3. Measurement of level of βLG in samples of whey receiving small increments in levels of heat treatment. Sigma grade V βLG, ■; unheated whey, ●; whey heated at 100°C for 10 min, ▲; 15 min, ◆; 20 min, ○; and 30 min, □.

Specificity

The possibility that samples of heated whey may cause non-specific inhibition in our assay was investigated. Samples of β-conglycinin, a soyabean protein, were diluted in phosphate-buffered saline (PBS) or in a 1:4 dilution of autoclaved whey (115°C, 30 min). These were then assayed by an ELISA of similar design to that for milk protein, using an antiserum to the specific antigen. Non-specific effects were not observed in the presence of heated whey (Fig. 4) which was therefore unlikely to

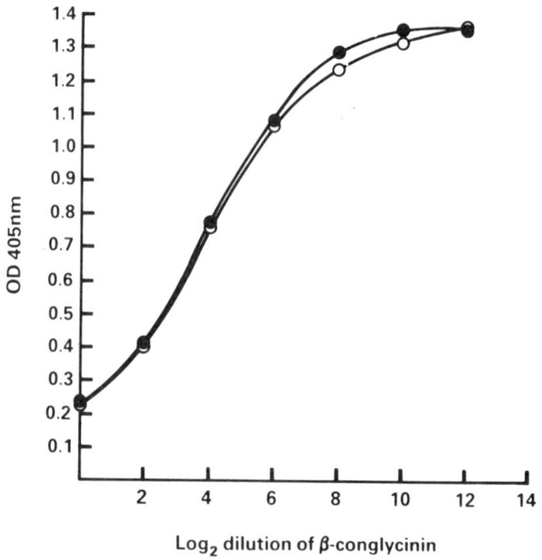

Fig. 4. Specificity of inhibition: effect of heated whey on assay of β-conglycinin. β-conglycinin diluted in PBS, ●; β-conglycinin diluted in a 1:4 dilution of autoclaved whey, ○.

have exerted a non-specific inhibitory effect on the milk protein ELISA.

Several heated whey samples contained a thick precipitate which might inhibit binding of specific antibody to the coated plates. This possibility was examined by allowing the precipitate to sediment during the overnight incubation of dilutions of the heated sample with anti-βLG antiserum. Aliquots (200 μl) were then transferred to the βLG-coated plate, either mixing before transfer to resuspend the precipitate, or without mixing to exclude the precipitate from further stages of the

Table 2
Effect of a Precipitate in the Competitive Inhibition ELISA for βLG

Whey sample	*Concentration of βLG (μg/ml)*	
	Precipitate present[a]	*Precipitate absent*[b]
1	0·41	0·56
2	0·79	0·85

[a]Precipitate resuspended before transfer to βLG-coated plate.
[b]Precipitate sedimented and supernatant transferred to βLG-coated plate.

assay. Table 2 shows that the presence of a precipitate in the samples had little effect on the values obtained.

Comparison with results *in vivo*

A detailed description of the effect of heat treatment on the antigenicity of cow's milk proteins has already been published (Kilshaw *et al.*, 1982; Heppell *et al.*, 1984). A comparison between results obtained by ELISA and by feeding experiments using guinea pigs is shown in Table 3.

All guinea pigs drinking pasteurised whey became sensitised for systemic anaphylaxis and produced serum antibodies to whey proteins. In contrast, heat treated whey failed to sensitise for anaphylaxis and evoked little or no antibody production. This reduction in sensitising capacity *in vivo* was accompanied by a corresponding reduction in the levels of antigenic whey protein detected *in vitro* by ELISA. Thus, once the relationship between results obtained *in vivo* and by ELISA has been established, the potential antigenicity *in vivo* of further samples of the protein can be predicted *in vitro*.

We are currently investigating the heating conditions required for the production of a hypoallergenic infant milk formula based on heat denatured whey. The number of samples tested in guinea pigs has been significantly reduced by using the ELISA to select for those samples likely to be weakly or non-sensitising *in vivo*.

These results must not be interpreted to indicate that the *in vivo* antigenicity of *all* heat treated proteins can be predicted by ELISA. Heated milk (121°C, 20 min) evoked lower levels of serum antibodies to α_{s_1}-casein than unheated milk (Kilshaw *et al.*, 1982). However, the level of antigenic α_{s_1}-casein detected by ELISA was 5-fold greater in the

Table 3
Comparison of Antigenicity Measured *in vitro* by ELISA with Results *in vivo**

			Pasteurised whey	*Whey heated 100°C 30 min*	*Whey heated 115°C 30 min*
In vivo	Fatal anaphylaxis after i.v. injection of fed preparation (no. affected/total)		4/4	0/10	0/5[a]
	Serum antibodies (mean IgG titres[b]) to:	βLG[c]	7·6 (6·5)	Negative (2·2)	Negative (0·5)
		αLA	0·2 (0·6)	Negative	Negative
		$BIgG_1$	7·0	Negative	Negative
		BSA	0·4	Negative	Negative
In vitro (by ELISA)[d]	Level of residual antigenic protein:	βLG	1 (1/2)	1/2048	1/8192 (<1 μg/ml)
		αLA	1	1/32	1/512
		$BIgG_1$	1 (1/2)	None detected	None detected
		BSA	1	<1/64	< 1/64

Where results of duplicate experiments differed, these are shown in brackets.
*Data from Heppell *et al.* (1984).
[a]Animals given i.v. injection of skimmed milk.
[b]Number of doubling dilutions from 1/20.
[c]βLG, β-lactoglobulin; α-lactalbumin; $BIgG_1$, bovine IgG_1; BSA, bovine serum albumin.
[d]Fraction to the nearest doubling dilution of level in unheated skimmed milk.

heated preparation (possibly due to changes in micellar structure during heating, resulting in increased accessibility of antigenic determinants to antibody). Thus for casein, the ELISA does not predict sensitising capacity *in vivo*.

In conclusion, these results show that the competitive inhibition ELISA can predict the sensitising capacity of heated whey proteins *in vivo*. The assay is simple, specific, reproducible and capable of detecting low levels of antigenic protein. The observation that assay curves for heated and unheated proteins differed in shape suggests that the residual antigenic proteins detected after heating were not immunologically identical to the unheated standards. This effect became more pronounced with increasing heat treatment. Consequently, the reliability of any serological assessment of residual undenatured protein may become questionable with increasing severity of heat treatment.

ACKNOWLEDGEMENT

I would like to acknowledge the contribution of Dr P. Kilshaw to this study.

REFERENCES

HEPPELL, L. M. J. & KILSHAW, P. J. (1982) Immune responses of guinea pigs to dietary protein. 1. Induction of tolerance by feeding with ovalbumin. *International Archives of Allergy & Applied Immunology*, **68**, 54–9.

HEPPELL, L. M. J., CANT, A. J. & KILSHAW, P. J. (1984) Reduction in the antigenicity of whey proteins by heat treatment; a possible strategy for producing a hypoallergenic infant milk formula. *British Journal of Nutrition*, **51**, 29–36.

KILSHAW, P. J., HEPPELL, L. M. J. & FORD, J. E. (1982) Effects of heat treatment of cow's milk and whey on the nutritional quality and antigenic properties. *Archives of Diseases in Childhood*, **57**, 842–7.

MCLAUGHLAN, P., ANDERSON, K. J., WIDDOWSON, E. M. & COOMBS, R. R. A. (1981) Effect of heat on the anaphylactic-sensitising capacity of cow's milk, goat's milk and various infant formulae fed to guinea pigs. *Archives of Diseases in Childhood*, **56**, 165–71.

VOLLER, A., BIDWELL, D. E. & BARTLETT, A. (1976) Enzyme immunoassays in diagnostic medicine. Theory and practice. *Bulletin of the World Health Organisation*, **53**, 55–65.

10

An Enzyme-linked Immunosorbent Assay for Amyloglucosidase in Beer

P. VAAG

Department of Biotechnology, Carlsberg Research Laboratory, Copenhagen, Denmark

ABBREVIATIONS

AMG	Amyloglucosidase
BSA	Bovine serum albumin
DEA	Diethanolamine
GI, GII	Two forms of amyloglucosidase from *Aspergillus niger*
IgG	Immunoglobulin G
PBS	Phosphate-buffered saline
PEG 6000	Polyethyleneglycol 6000
PNPP	*p*-Nitrophenylphosphate

INTRODUCTION

Amyloglucosidase (EC 3.2.1.3) is an enzyme capable of hydrolysing both α-1,4 and α-1,6 glucosidic linkages of starch, glycogen and glucooligosaccharides, releasing free D-glucose. Unlike amylases, which only attack the α-1,4 linkages and thus leave undegradable limit dextrins containing α-1,4 and α-1,6 linkages after exhaustive action on starch, AMG is able to convert starch almost completely to glucose. This has led to industrial application of AMG in the production of beer types with ordinary alcohol levels, but lower calorie content.

In the production of ordinary beer, the limit dextrins present in wort after the action of the enzymes released during malting and mashing are not further degraded during fermentation. They are thus present in the finished beer, where, together with the alcohol, they contribute to the total calorie content. By addition of AMG during fermentation, a large part of the limit dextrins are hydrolysed, and free glucose is thus made available for the yeast for growth and ethanol production. A beer produced with AMG will, therefore, contain less calories from dextrins.

At present, a variety of AMG preparations produced from cultures of *Aspergillus niger* are being used for production of low calorie beers. However, the extreme heat stability of the enzyme — a large part of the activity survives ordinary pasteurisation — gives rise to problems in breweries making both ordinary and low calorie types of beer. If a small amount of AMG beer is accidently mixed with ordinary beer, free glucose will be released during storage and will add an unacceptable sweet taste to the beer.

Beer samples have hitherto been tested routinely for AMG by comparison of the glucose content of a 'hot' sample, incubated at 60°C for 1·5 h (or, in cases of doubt, for up to 24 h), and a 'cold' sample, kept in an ice bath during the same period. Glucose in the samples is estimated by colour development on paper sticks impregnated with a glucose oxidase system. This method is very simple to perform, but has several drawbacks. One is that it can be difficult to determine differences in colour intensity, and the sensitivity is low — the presence of AMG below approximately 2% of the level normally found in low calorie beers is hardly detected. Another is that the method is not specific for AMG, as the glucose development theoretically can be caused by other enzymes (i.e. β-glucanase degrading β-glucans). There is thus a need for a more sensitive and less ambiguous assay.

Investigations were therefore carried out concerning the possibilities of establishing a specific ELISA procedure, which could meet the following requirements: (1) the limit of detection of AMG should be about 0·1% of the level normally found in low calorie beers, (2) the total assay time should not exceed 4–5 h, and (3) analysis results should eventually be easily estimated by eye.

MATERIALS AND METHODS

Special Reagents and Equipment

Amyloglucosidase preparations for use in production of low-calorie beers were Novo AMG 150 LP (Novo, Denmark); Ambazyme (ABM, England); Amylo 100 (Biocon, England); and Diazyme L-150 (Miles,

USA), α-Amylases from *Aspergillus oryzae* and from barley malt, bovine serum albumin (BSA, type V), alkaline phosphatase (type VII-S) and *p*-nitrophenylphosphate tablets were from Sigma, USA. Protein A-Sepharose CL-4B was from Pharmacia Fine Chemicals AB, Sweden. Agarose was type HSB from Litex, Denmark. Soluble starch, Tween 20 and the proteases papain, ficin and bromelin were from Merck, West Germany. Polyethyleneglycol 6000 was from BDH, England. Other chemicals were analytical grade.

A low calorie beer produced with AMG and an ordinary beer type without AMG were obtained.

The ELISA procedure was performed in disposable polystyrene cuvettes, arranged in blocks of nine, from Labsystems Oy, Finland. Adjustable nine-channel pipettes, a thermostated incubator and a nine-channel spectrophotometer based on vertical measurement (FP-901), all fitted for the above mentioned cuvette blocks, were obtained from the same company.

Antibodies

Crude antiserum, containing polyspecific antibodies directed against AMG and other antigens present in commercial preparations of AMG for use in breweries, was prepared by Dako, Denmark, by immunisation of three rabbits with a mixture of the commercial preparations mentioned above.

The sera obtained were evaluated with various precipitation-in-gel techniques (crossed immunoelectrophoresis, double diffusion in two dimensions, etc.) as described by Weeke (1973) and Vaag & Gibbons (1982). Starch-degrading activities (amyloglucosidase, amylases) retained by immunoprecipitation in agarose gels were demonstrated by incubating the washed and pressed gels in a bath of 1% (w/v) soluble starch in 100 mM phosphate buffer, pH 4·0, at 50°C for 30 min as described by Hejgaard (1976). After development in an I_2/KI bath, starch-degrading activity appeared as white or light violet peaks on a blue background. Plates can be destained and later stained for protein with Coomassie Blue.

Immunoglobulin G was purified from crude serum by affinity chromatography on a Protein A-Sepharose CL-4B column as described by the supplier. Protein content in the purified preparation was determined by the method of Lowry *et al.* (1951).

Preparation of Conjugate

A conjugate between Protein A-purified IgG and alkaline phosphatase was produced by the one-step glutaraldehyde method as described by

O'Sullivan & Marks (1981). The mixture of enzyme–antibody conjugate, free or intra-conjugated (polymerised) enzyme, and free or intra-conjugated antibody obtained by this procedure was used directly in the experiments described in the following text, without further purification.

ELISA-procedure for Detection of AMG

A non-competitive type of assay, the double antibody sandwich method, was chosen for detection of AMG in beer. This type of assay involves four steps: (1) coating polystyrene cuvettes with antibody, (2) incubation with the solution to be tested for antigen, i.e. beer, (3) incubation with enzyme–antibody conjugate, and (4) incubation with a substrate for the enzyme part of the conjugate. Between steps, cuvettes are washed thoroughly to remove unbound reactants.

The four steps were performed as follows:

(1) Purified anti-AMG IgG was diluted in phosphate-buffered saline (PBS_{10}: 10 mM sodium phosphate, 150 mM sodium chloride, pH 7·2) to a concentration of 1, 5 or 20 μg/ml. To each cuvette were added 200 μl of IgG solution, after which the cuvettes were left for 20–30 h at 4°C. After three washes with 400 μl PBS/Tween (PBS_{10} with 0·01% Tween 20), cuvettes were either used immediately or washed once more with distilled water and stored in a cold room in boxes containing silica gel.

(2) The beer samples were decarbonised by pouring the samples several times from one beaker to another. Their pH values were then adjusted to approximately 7·0 by mixing four volumes of beer with one volume of a five times stock solution of PBS_{50} (5 × stock of PBS_{50}: 250 mM sodium phosphate, 750 mM sodium chloride, pH 7·2). It is important to degas the beer samples properly and to employ a rather high buffer strength for pH adjustment, as slow decarbonisation during incubation in the cuvettes will otherwise raise the pH of the samples and thus produce irregularities. To each cuvette, 200 μl of pH-adjusted beer were added. Incubation was performed at room temperature for 1–4 h, after which the cuvettes were again washed three times with PBS/Tween.

(3) 200 μl of enzyme-labelled antibody conjugate, diluted 1:500 in PBS_{10} containing various additives, were then incubated in the cuvettes, either at room temperature for 2 h or at 4°C for 12–20 h. Following incubation, cuvettes were washed as before.

(4) 200 μl of a solution of *p*-nitrophenylphosphate in DEA-buffer (2 mg/ml) (DEA-buffer: 1·0 M diethanolamine, 0·5 mM $MgCl_2$, adjusted to pH 10·0 with HCl) were added and incubated at 37°C. The reaction was

stopped after 20 or 30 min by addition of 200 μl 1 M NaOH, and the absorbance at 405 nm was read in the nine-channel spectrophotometer against a blank of 200 μl substrate solution and 200 μl NaOH.

RESULTS

Evaluation of Antisera

Crossed immunoelectrophoresis revealed that sera from all three rabbits contained antibodies towards both AMG and a few other antigens (Figs 1(a) and (b)). The two main peaks showing partial immunological identity, together with a minor peak, are thought to represent the two related forms of *Aspergillus niger* amyloglucosidase, GI and GII, which have previously been demonstrated to cross-react immunologically (Lineback *et al.*, 1969; Pazur *et al.*, 1971). Svensson *et al.* (1981) have reported GI to give a major band at pI 4·0 and a minor one at pI 3·6 in preparative flat bed isoelectric focusing, while GII gives one band at pI 4·2. The two peaks closest to the anode are thus presumed to be the two forms of GI, differing in isoelectric point, while the dominant peak closest to the cathode is presumed to be GII. Apart from these three peaks, one of the antigenic contaminants of commercial AMG preparations exhibited a strong starch-degrading activity and is presumed to be an *Aspergillus niger* amylase. In agreement with this, a faint cross-reaction was observed between all three sera and α-amylase from the related species *Aspergillus oryzae*, when tested by the method of double diffusion in two dimensions (Fig. 2).

The relative content of the antigens varied widely from one brand of AMG to another, but, as judged from the intensity of the protein-stained immunoelectrophoresis plates, GI/GII were by far the dominating antigens in all brands.

No cross-reactions were detected by double diffusion in two dimensions between any of the sera and barley malt α-amylase or the plant proteases (papain, ficin, bromelin) commonly employed for beer stabilisation. Likewise, no immunoprecipitates could be detected after crossed immunoelectrophoresis performed with beer samples known to be devoid of AMG, while beer samples containing AMG gave rise to several peaks, of which two could be identified as amyloglucosidase and amylase (Figs 3(a) and (b)).

In spite of the presence of antibodies directed against antigens other than AMG (GI/GII), it was decided to work out an ELISA procedure,

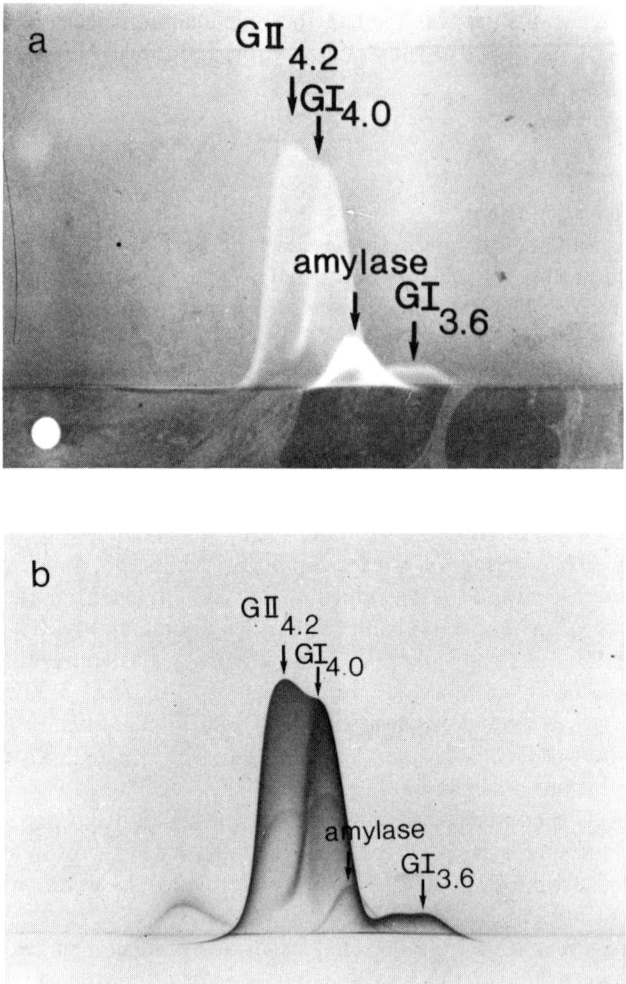

Fig. 1. Crossed immunoelectrophoresis of a commercial AMG preparation, Ambazyme, versus polyspecific antiserum. Ambazyme (13·5 μg) was applied in the antigen well. The second dimension antibody-gel contained 10 μl crude serum/cm^2. First dimension: anode at right; second dimension: anode at top. The marked peaks are presumed to represent GII, the two forms of GI differing in pI, and amylase. (a) The plate stained for starch-degrading activity with starch–iodine, (b) the same plate destained and stained for protein with Coomassie Blue.

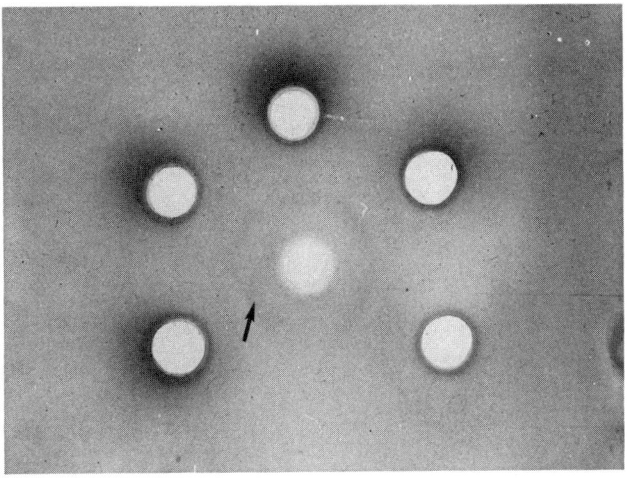

Fig. 2. Double diffusion in two dimensions between *Aspergillus oryzae* α-amylase and polyspecific antisera. α-Amylase (200 μg) was applied in the central well. To the peripheral wells were added 20 μl of crude serum obtained from various rabbits and bleedings. A faint precipitate, marked by arrow, is observed.

based on the polyspecific sera obtained, in order to determine whether a usable assay as regards the required sensitivity and speed could be produced. The effect of various assay conditions on the ELISA response from mixtures of low calorie beer and ordinary beer were therefore studied as outlined below.

Effect of pH of Beer Samples

Initially, ELISA assays were performed with mixtures of low calorie beer and ordinary beer adjusted to various pH levels (Fig. 4). Obviously, pH values between approximately 6·5–8·0 were beneficial for the binding between the IgG attached to the cuvettes and AMG in the samples (Fig. 4). All other experiments were thus performed with samples adjusted to approximately pH 7·0 with PBS.

Effect of Concentration of IgG in Coating Solution

Cuvettes were coated with solutions of IgG at a concentration of either 1, 5 or 20 μg/ml and incubated with beer samples containing 0–50% low calorie beer (Fig. 5). Of these alternatives, too low a sensitivity was obtained when coating with only 1 μg/ml. Coating with 5 or 20 μg/ml

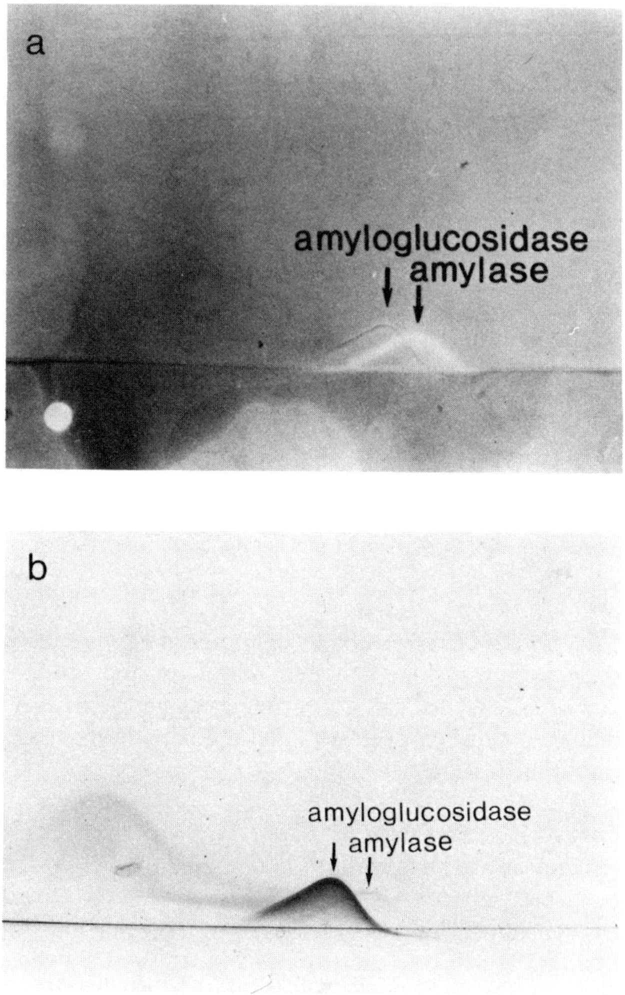

Fig. 3. Crossed immunoelectrophoresis of a low calorie beer versus polyspecific antiserum. The beer sample was concentrated by precipitation with 70% $(NH_4)_2SO_4$, dialysis and lyophilisation as described by Vaag & Gibbons (1982). Lyophilisate (250 μg) was applied in the antigen well. The second dimension gel contained 10 μl crude serum/cm^2. Peaks representing AMG and amylase are indicated. (a) Stained for starch-degrading activity, (b) stained for protein.

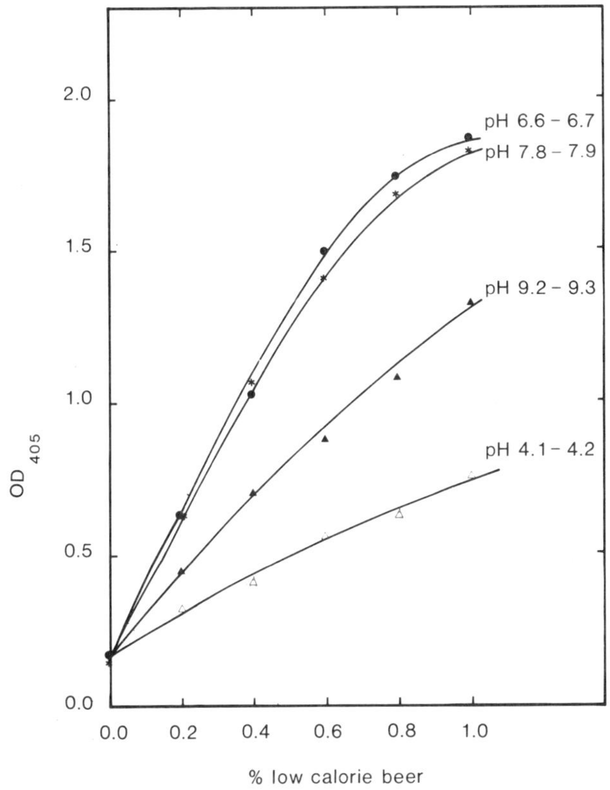

Fig. 4. Effect of pH during incubation with beer samples. Assay conditions: (1) cuvettes were coated with 5 µg IgG/ml, (2) beer samples, adjusted to various pH values, were incubated at room temperature for 3·5 h, (3) conjugate was diluted 1:500 in PBS_{10} containing 1% BSA and incubated at 4°C for 18 h, and (4) PNPP was incubated for 30 min at 37°C.

gave detection limits at about 0·1–0·2% low calorie beer, and straight standard curves in the region 0·0–1·0% low calorie beer could be obtained when plotted on linear axes (Fig. 6). Coating with 20 µg/ml gave somewhat higher absorbances, but the ratios between absorbances for 1·0% and 0·0% low calorie beer were almost equal.

Effect of Incubation Time with Beer Samples

Varying the time of incubation with the beer samples from 1–3 h (Fig. 7) gave progressively higher absorbances for mixtures with low calorie beer

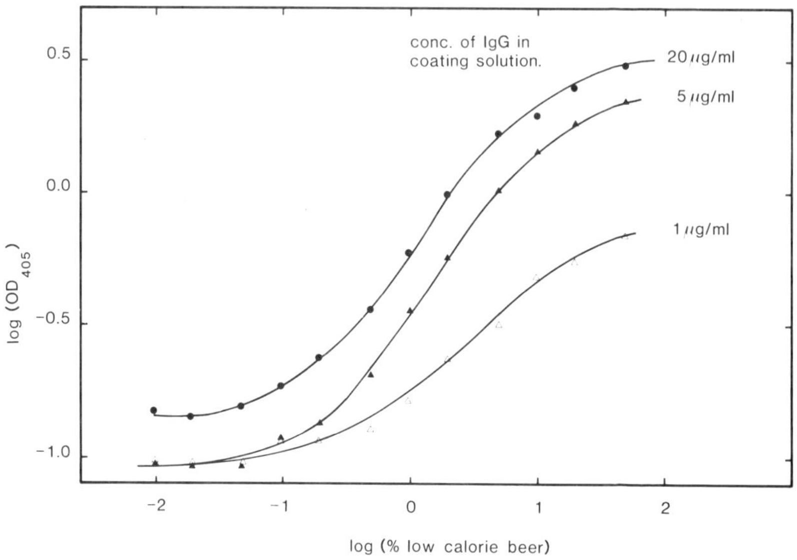

Fig. 5. Effect of concentration of IgG in coating solution. Assay conditions: (1) cuvettes were coated with either 1, 5 or 20 μg IgG/ml, (2) beer samples, adjusted to pH 7·0 with PBS, were incubated for 2·5 h at room temperature, (3) conjugate, diluted 1:500 in PBS with 1% BSA, was incubated at 4°C for 12 h, and (4) PNPP was incubated for 20 min at 37°C.

while the absorbance for beer samples without AMG was unaffected. Higher sensitivity can thus be obtained by prolongation of the incubation time with the beer samples; but as it is essential for the practical applicability of the ELISA procedure that the whole assay can be performed as quickly as possible, long incubation times are not realistic.

Effect of Various Additives to the Conjugate Dilution Buffer

In an effort to reduce the time required for incubation with enzyme–antibody conjugate, the effects of PEG 6000, which is known to enhance the reaction between antigen and antibody (Hellsing, 1973), and Tween 20 were investigated. Conjugate was incubated at room temperature for 2 h, having been diluted in PBS_{10} to which had been added either 1% BSA alone, or 1% BSA in combination with PEG 6000 (1, 2 or 3%) and Tween 20 (0 or 0·01%).

The addition of PEG 6000 to the conjugate dilution buffer elevated the absorbances obtained for samples with AMG in all experiments.

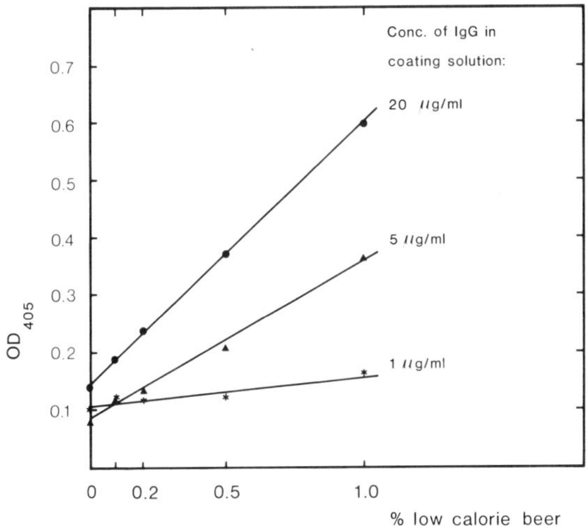

Fig. 6. Effect of concentration of IgG in coating solution for AMG levels equivalent to 0·0–1·0% low calorie beer. Assay conditions: as outlined for Fig. 5.

However, high concentration of PEG 6000 also elevated the absorbances for samples devoid of AMG considerably, especially when Tween 20 was also added (Fig. 8). Attempts to employ high concentration of PEG 6000 in combination with very short incubation times were not successful, as blank levels were unduly high also in these cases.

However, a fair compromise between speed and sensitivity seems to be obtainable by addition of 2% PEG 6000 to the conjugate dilution buffer.

Effect of Various Other Assay Conditions

Addition of PEG 6000 (1 or 2%) to beer samples prior to incubation had almost no effect on the absorbances and thus could not reduce the total assay time. Raising the temperature to 37°C during incubation of beer samples and/or conjugate gave rise to rather high blank levels and great variation between triplicate assays.

DISCUSSION

The requirements for an ELISA procedure for detection of AMG in beer outlined in the introduction can be fulfilled if assays are performed in the following way: beer samples, adjusted to approximately pH 7·0, are

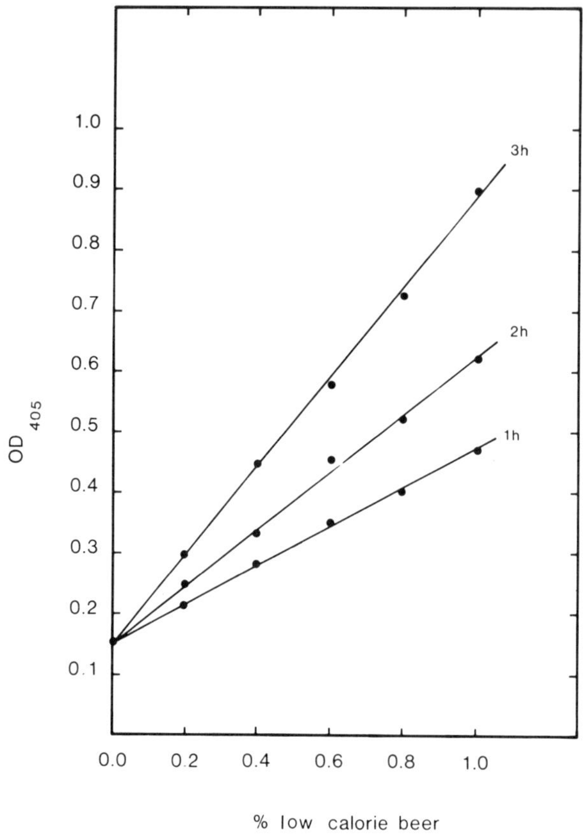

Fig. 7. Effect of incubation time with beer samples. Assay conditions: (1) cuvettes were coated with 5 μg of IgG/ml, (2) beer samples, adjusted to pH 7·0 with PBS, were incubated for 1, 2 or 3 h at room temperature, (3) conjugate, diluted 1:500 in PBS with 1% BSA, was incubated at 4°C for 19 h, (4) PNPP was incubated at 37°C for 20 min.

incubated for 2 h at room temperature in polystyrene cuvettes, coated with 5 μg anti-AMG IgG/ml. After washing, this is followed by incubation with alkaline phosphatase–anti-AMG IgG conjugate, diluted 1:500 in PBS_{10} containing 1% BSA and 2% PEG 6000, for 2 h at room temperature. Following a further washing, the cuvettes are incubated with *p*-nitrophenylphosphate, 2 mg/ml in DEA-buffer, for 30 min at 37°C. A ratio of 7·4 between the absorbances of beer samples equivalent to 1% of low calorie beer and beer samples devoid of AMG is thus obtained.

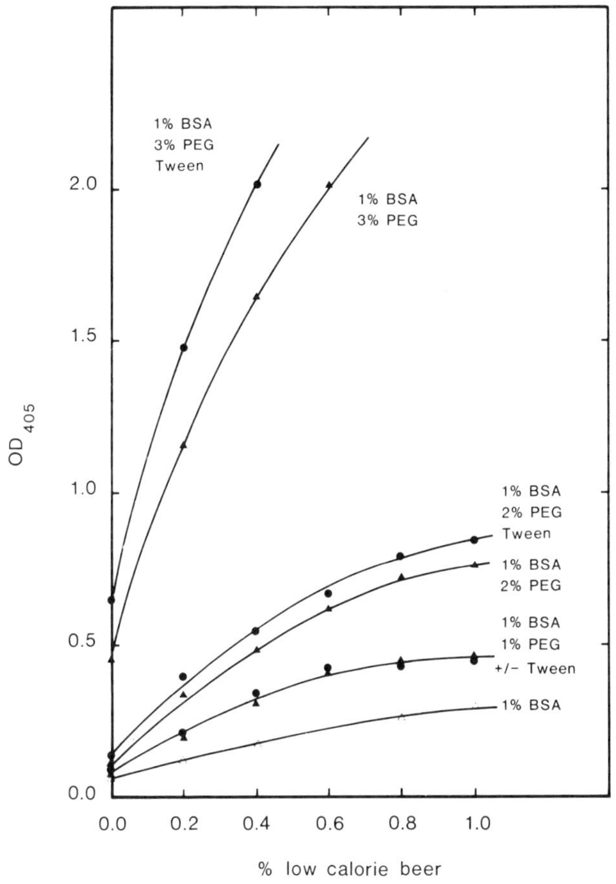

Fig. 8. Effect of addition of various compounds to the conjugate dilution buffer. Assay conditions: (1) cuvettes were coated with 5 μg IgG/ml, (2) beer samples, adjusted to pH 7·0 with PBS, were incubated at room temperature for 2 h, (3) conjugate, diluted 1:500 in PBS containing various additives as indicated on the figure, was incubated for 2 h at room temperature, and (4) PNPP was incubated at 37°C for 20 min.

This is comparable to the ratio obtained when incubating conjugate, diluted 1:500 in PBS_{10} with 1% BSA alone, at 4°C overnight. However, it must be noted that the decrease in total assay time is obtained at the expense of the linear correlation between content of AMG and ELISA response (compare Figs 8 and 6), but as the deviation is not great, this seems acceptable.

The possibility of achieving higher sensitivity and/or shorter total assay time by using purified conjugate preparations, where enzyme–antibody conjugate has been separated from unconjugated and polymerised reactants, is currently being investigated.

The experiments have been carried out with polyspecific antisera containing antibodies against not only AMG, but also amylase and a few other antigens. This means that the observed variations between commercial AMG brands, as regards the relative content of the different antigens, will bring about variations in the ELISA response obtained from low calorie beers produced with the same amount of AMG, but from different brands. However, as the amount of antigens other than AMG in the tested samples is small, their contribution to the ELISA response is also small, and it can, of course, be totally eliminated by the use of antibodies directed only against AMG. Therefore, antibodies against GI (which will probably cross-react with GII) are being sought.

A more complex problem is posed by the different enzymatic activities of GI and GII, GI being respectively 10–30 and 1·3–1·5 times more active than GII against raw starch and soluble polysaccharides from various sources (Svensson *et al.*, 1981). However, once the relationship between the ELISA response and the combined enzymatic activities of GI and GII is established for a given brand of AMG (and a given batch of this brand), the difference in enzyme activity between the two cross-reacting forms of AMG will be of no importance for the practical application of the ELISA procedure in breweries' control laboratories. The relationship between content of GI/GII, response in ELISA assays and actual activity on dextrins in beer will be the subject of a further study.

ACKNOWLEDGEMENTS

The skilled technical assistance of Ms A. G. Hansen as well as the support by Dr L. Munck, Head of the Department of Biotechnology, Carlsberg Research Laboratory, is gratefully acknowledged by the author.

REFERENCES

HEJGAARD, J. (1976) Free and protein-bound β-amylases of barley grain. Characterization by two-dimensional immunoelectrophoresis. *Physiologia Plantarum*, **38**, 293–9.

HELLSING, K. (1973) The effects of different polymers for enhancement of the antigen–antibody reaction as measured with nephelometry. *Protides of Biological Fluids*, **21**, 579.

LINEBACK, D. R., RUSSELL, I. J. & RASMUSSEN, C. (1969) Two forms of the glucoamylase of *Aspergillus niger*. *Archives of Biochemistry and Biophysics*, **134**, 539–53.

LOWRY, O. H., ROSEBROUGH, N. J., FARR, L. & RANDALL, R. J. (1951) Protein measurement with the Folin phenol reagent. *Journal of Biological Chemistry*, **193**, 265–75.

O'SULLIVAN, M. J. & MARKS, V. (1981) Methods for the preparation of enzyme–antibody conjugates for use in enzyme immunoassay. In: *Immunochemical Techniques (Part B)*, Langone, J. J. & Van Vunakis, H. (eds), *Methods in Enzymology*, Vol. 73, Academic Press, New York, pp. 147–66.

PAZUR, J. H., KNULL, H. R. & CEPURE, A. (1971) Glycoenzymes: structure and properties of the two forms of glucoamylase from *Aspergillus niger*. *Carbohydrate Research*, **20**, 83–96.

SVENSSON, B., PEDERSEN, T. G., SVENDSEN, I., SAKAI, T. & OTTESEN, M. (1981) Characterization of two forms of glucoamylase from *Aspergillus niger*. *Carlsberg Research Communications*, **47**, 55–69.

VAAG, P. & GIBBONS, G. C. (1982) Praktische Verfahren für die immunologische Bieranalytik. *Brauwissenschaft*, **35**, 241–8.

WEEKE, B. (1973) Crossed immunoelectrophoresis. In: *A Manual of Quantitative Immunoelectrophoresis, Methods and Applications*, Axelsen, N. H., Krøll, J. & Weeke, B. (eds), Universitetsforlaget, Oslo, Norway, pp. 47–56.

11

Application of Enzyme Immunoassay Techniques for the Estimation of Staphylococcal Enterotoxins in Foods

P. D. PATEL

Leatherhead Food Research Association, Leatherhead, Surrey, UK

INTRODUCTION

Normally, foods implicated in staphylococcal food-poisoning outbreaks contain small amounts of enterotoxins (Minor & Marth, 1976). Enterotoxins A and D are most frequently encountered in food-poisoning outbreaks (Holbrook & Baird-Parker, 1975). Since enterotoxin A usually occurred in concentrations of ≤1 μg/100 g of food it was suggested that the sensitivity of any enterotoxin assay technique should be between 0·125–0·25 μg of enterotoxin per 100 g of food sample (Reiser *et al.*, 1974).

The procedures currently used for the detection and estimation of staphylococcal enterotoxins in foods are complex and lengthy but have been demonstrated to perform satisfactorily in many laboratories. The procedures normally involved are the extraction of enterotoxin from the food (2–3 days) followed by detection of the enterotoxin by the micro-slide double-immunodiffusion technique (a further 2–3 days) (Crowle, 1958). Thus a result may be obtained within a week but frequently takes longer (Holbrook & Baird-Parker, 1975). The sensitivity of double-immunodiffusion techniques for detecting enterotoxins in foods varies from 0·1 to 2 μg enterotoxin per 100 g of food sample (Niskanen, 1977).

In contrast, radioimmunoassay (RIA) techniques are rapid in regard to the total time taken for estimation of enterotoxins in foods (≤1 day) and have been used in several laboratories (Miller *et al.*, 1978; Areson *et al.*, 1980; Bergdoll & Reiser, 1980). Although RIA techniques are highly sensitive (detection limit 1–10 ng enterotoxin/ml) and reproducible, their use is restricted by the requirement for radioisotopes and expensive reagents and instruments.

Over the last decade enzyme immunoassay (EIA) techniques have been developed for antigens and antibodies (Weemen & Schuurs, 1971; Dubois-Dalcq *et al.*, 1977; Voller & Bidwell, 1980) and have been the subject of several reviews (Wisdom, 1976; O'Sullivan *et al.*, 1979) and books (Voller *et al.*, 1979; Voller & Bidwell, 1980; Maggio, 1981). Like RIA, EIA techniques are rapid and highly sensitive. A common variant of the technique is referred to as enzyme-linked immunosorbent assay — ELISA. The ELISA technique was first reported for the determination of rabbit immunoglobulin G by Engvall & Perlmann (1971). Since then many variants of enzyme immunoassay techniques have been reported with applications in many areas such as in clinical and veterinary medicine and in the agricultural and food industries (Voller & Bidwell, 1980; Chapter 1).

There are broadly three types of ELISA technique which have been applied to the estimation of staphylococcal enterotoxins in enterotoxigenic culture filtrates and/or foods.

COMPETITIVE ELISA TECHNIQUES

The antibodies (Ab) against enterotoxins are adsorbed to a solid phase, e.g. polystyrene tube, plate or ball. A mixture of toxin (T) in the sample to be determined and the conjugate (T–Enz) of toxin covalently coupled to an enzyme (e.g. horseradish peroxidase) is incubated with the antibody-adsorbed solid phase. Unreacted reagents are removed by washing and the enzyme activity of the conjugate retained on the solid phase is determined. The process is summarised in Fig. 1. This enzyme activity is inversely related to the initial amount of toxin present in the sample (Fig. 2).

Competitive ELISAs have been reported for staphylococcal enterotoxins B (Simon & Terplan, 1977) and A, B and C (Stiffler-Rosenberg & Fey, 1978; Lenz *et al.*, 1983). Velan & Halmann (1978) reported a chemiluminescence immunoassay for staphylococcal enterotoxin B. The assay was basically a competitive ELISA in which the activity of horseradish peroxidase was measured by a chemiluminescence method using pyrogallol and H_2O_2 instead of a colorimetric estimation. A conventional enzyme immunoassay for enterotoxin A was reported by Kauffman (1980). In this technique antibodies were not adsorbed to a

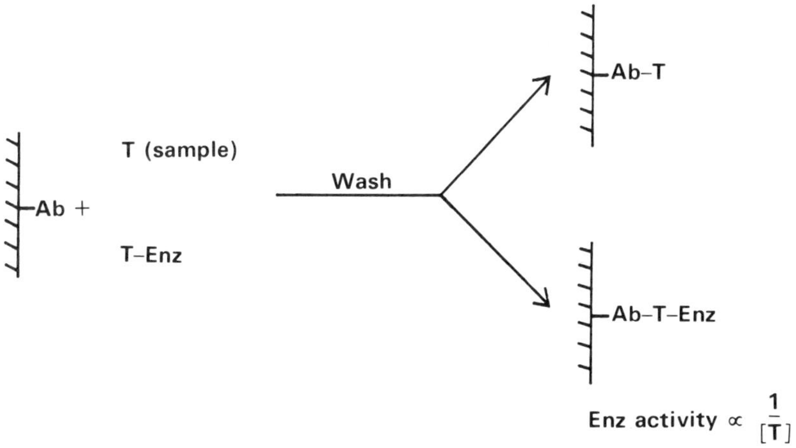

Fig. 1. Competitive ELISA technique: Ab, antibody; T, toxin; T–Enz, enzyme-labelled toxin.

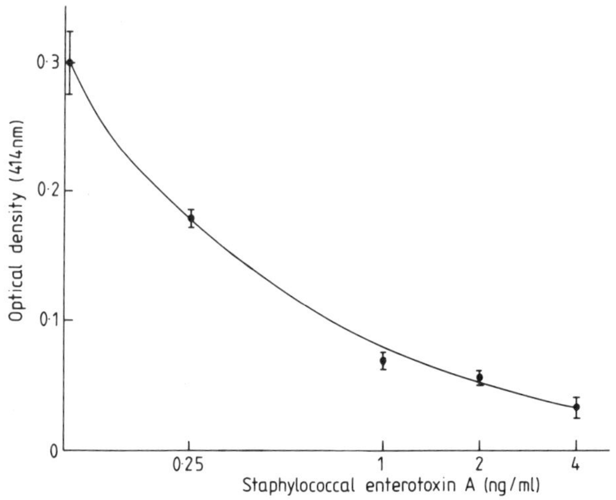

Fig. 2. Use of peroxidase in the ELISA of staphylococcal enterotoxin A (reproduced with permission from Kuo & Silverman, 1980).

solid phase. Instead, anti-enterotoxin A antiserum was incubated with a mixture of the sample or standard containing enterotoxin A and a conjugate of enterotoxin A coupled to alkaline phosphatase. After incubation, the immune complex (antibody–enterotoxin–enzyme) was separated from the unbound conjugate using staphylococcal cells containing Protein A on the cell surface. Protein A behaves like a second antibody in that it binds to the Fc portion of IgG immunoglobulin molecules of most species, leaving the two Fab portions free to interact with specific antigens (Endresen, 1978; Goding, 1978; Lindmark, 1982). The immune complexes adsorbed to staphylococcal cells can be readily separated by centrifugation (e.g. 2000 g for 5 min). Enzyme activity of the immune complexes will then be inversely related to the initial amount of toxin present in the sample.

NON-COMPETITIVE TECHNIQUES

(a) Direct Sandwich Methods

'Sandwich' methods, sometimes referred to as two-site assays, can only be used for antigens with at least two antigenic determinants. In this method, antibodies against enterotoxins are adsorbed to a solid phase as in the competitive ELISA technique. A sample containing toxin is incubated with the antibody to determinant A adsorbed to a solid phase. The unadsorbed toxin is removed by washing the solid phase and a conjugate of the specific antibody to determinant B, covalently coupled to enzyme (Ab–Enz), is added. The unbound conjugate is washed away and enzyme activity of the bound conjugate is determined (see Fig. 3). The enzyme activity is then directly related to the amount of toxin present in the sample (Fig. 4).

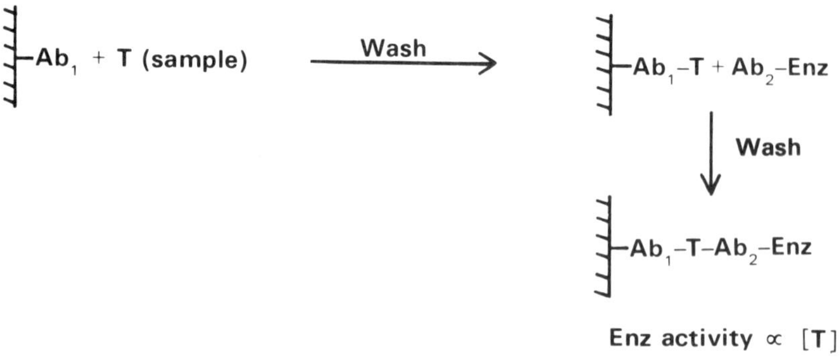

Fig. 3. Non-competitive direct sandwich ELISA technique: T, toxin; Ab_1, antibody to first antigenic determinant on toxin; Ab_2–Enz, enzyme-labelled antibody to second antigenic determinant on toxin.

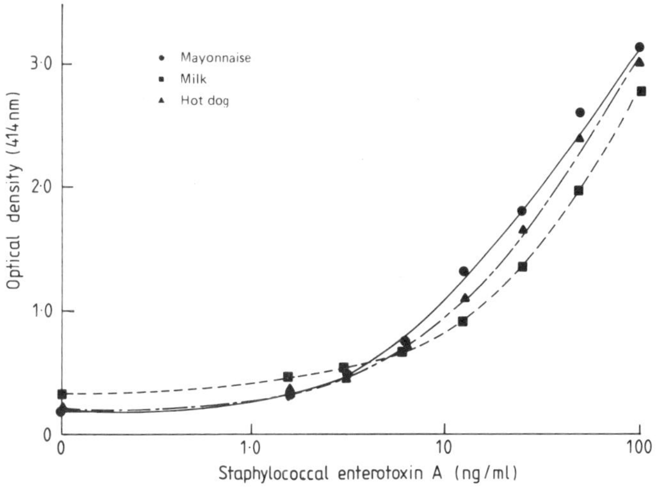

Fig. 4. Standard curves illustrating detection of staphylococcal enterotoxin A in food extracts (reproduced with permission from Saunders & Bartlett, 1977).

The direct sandwich ELISA techniques reported for various types of enterotoxins include enterotoxin A (Saunders & Bartlett, 1977; Kuo & Silverman, 1980; Ackerman & Chesbro, 1981), B (Notermans *et al.*, 1978; Koper *et al.*, 1980), A–C (Fey *et al.*, 1982) and A–E (Freed *et al.*, 1982). De Jong (1983) reported a non-competitive ELISA in which the antigen (i.e. enterotoxin A) was detected by using a conjugate of rabbit antibody specific for enterotoxin A and goat anti-rabbit IgG-alkaline phosphatase.

(b) Indirect Sandwich Methods

Antibodies (e.g. goat anti-enterotoxin) are adsorbed to a solid phase and, following washing, a sample containing the toxin is then added. After incubation the unadsorbed material is removed by washing and further enterotoxin antibodies, directed to a different antigenic determinant on the molecule and produced in a different animal species (e.g. rabbit anti-enterotoxin), are added. Following incubation and washing of immobilised antigen–antibody complexes on the solid phase, an enzyme conjugate of antibody, against the IgG of the second animal species, is added (Ab–Enz, e.g. goat anti-rabbit IgG-alkaline phosphatase). The unbound conjugate is separated from the bound material by washing the solid phase and the enzyme activity of the bound conjugate is then determined. The

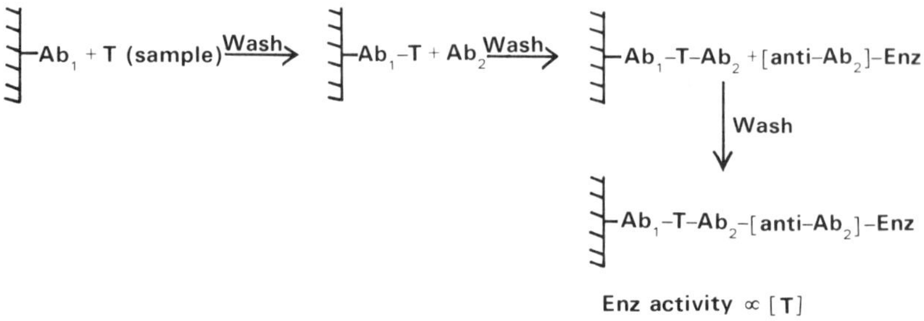

Fig. 5. Indirect sandwich ELISA technique: T, toxin; Ab$_1$, antibody to first antigenic determinant on toxin; Ab$_2$, antibody to second antigenic determinant on toxin raised in a different species; [Anti-Ab$_2$]–Enz, enzyme-labelled second antibody to IgG of species in which antibody to second determinant was raised.

enzyme activity is directly proportional to the initial concentration of toxin present in the sample (Fig. 5).

Indirect sandwich type of ELISA techniques have been reported for staphylococcal enterotoxins A, B and C (Berdal *et al.*, 1981; Olsvik *et al.*, 1982). Recently Fey *et al.* (1982) have reported a sandwich technique for staphylococcal enterotoxins in which instead of an anti-animal species antibody-coupled enzyme they have used Protein A coupled to alkaline phosphatase. Protein A binds to the Fc fragments of the IgG of a number of different species, but not those of sheep and goats.

Foods highly contaminated with *Staphylococcus aureus* may contain extracellular nontoxic staphylococcal Protein A (Koper *et al.*, 1980; Fey *et al.*, 1982). The presence of Protein A results in significant interference in ELISA for staphylococcal enterotoxins leading to reduced binding of the enzyme conjugate and thus an underestimate of the toxin content (Koper *et al.*, 1980; Berdal *et al.*, 1981; Freed *et al.*, 1982). The interference of Protein A has been completely eliminated by using $F(ab)'_2$ fragments instead of whole IgG molecules (Koper *et al.*, 1980). Alternatively, antibodies produced in sheep and goats may be used since Protein A has a lower affinity for the Fc fragments of the IgGs of these species compared to that of rabbit (Van der Ouderaa & Haas, 1981). Although food extracts may not contain large concentrations of Protein A, it has been suggested that a screen for Protein A interference be included in ELISA techniques applied to foods (Freed *et al.*, 1982).

ELISA is a heterogeneous system because a separation of antibody bound enzyme conjugate from free enzyme-labelled material is required. An alternative to ELISA was the development of a homogeneous enzyme-multiplied immunoassay technique (EMIT) which did not require separation of free from bound enzyme conjugate (Morita & Woodburn, 1978). The basis of EMIT is the inhibition of the activity of an enzyme coupled to enterotoxin when the latter is complexed with its antibody by the antibody obstructing the substrate binding site on the enzyme. This obstruction is removed by the presence of native enterotoxin. However, the results obtained by Morita & Woodburn (1978) could not be reproduced (Patel, 1982). Detailed investigation of EMIT by using a range of electrophoretic techniques strongly suggested that the basic requirement of the technique, the blockage of the enzyme activity of an enzyme–toxin conjugate by enterotoxin B antiserum, had not been achieved.

In the solid phased ELISA and RIA techniques, either the antigens or the antibodies are adsorbed non-covalently to solid surfaces such as polystyrene tubes, balls, beads or microtitre plates. In these techniques the separation of free label from the bound label is achieved either by washing and aspiration (e.g. tubes and plates) or centrifugation (e.g. beads). An alternative to the plastic solid surface was the development of magnetic beads, act-Magnogel AcA-44 (supplied by LKB Ltd, Bromma, Sweden), consisting of polyacrylamide–agarose gel which contains iron oxide in the matrix. The beads are activated with glutaraldehyde and therefore proteins (e.g. antibodies) with free amino groups can be covalently coupled to the gel. Since the beads can be separated from an aqueous phase simply by application of a magnetic field, the separation of unbound enzyme from the bound label can be accomplished rapidly. The magnetic beads have been used in novel techniques known as magnetic enzyme immunoassays (MEIA) for estimating immunoglobulins (Guesdon & Avrameas, 1977; Guesdon *et al.*, 1978*a*,*b*) and fractionation of lymphoid cells (Antoine *et al.*, 1978).

Recently we have developed a rapid (i.e. <7 h) non-competitive MEIA for the estimation of staphylococcal enterotoxin B IgG antibodies in both normal immune serum and an affinity purified IgG fraction (Patel *et al.*, 1983). An inhibition version of MEIA has also been developed (Patel & Gibbs, 1983) for the estimation of staphylococcal enterotoxin B (see Fig. 6).

The enterotoxin (T) in the sample to be determined is titrated against specific antibodies (produced in rabbit) present in excess to yield antibody–toxin (Ab–T) complexes. The residual free antibody is back

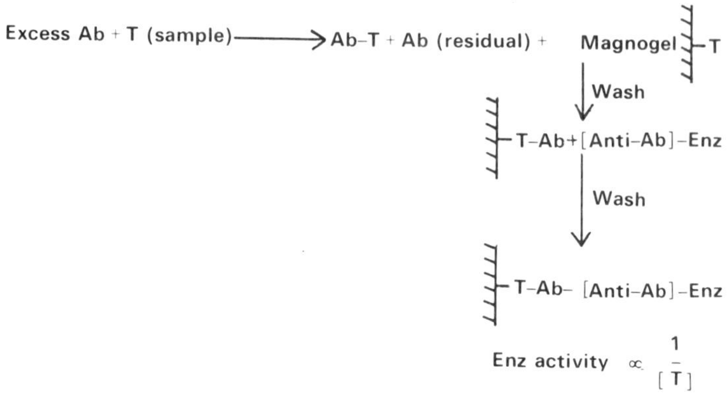

Fig. 6. Magnetic enzyme immunometric assay (MEIA) technique: T, enterotoxin; Ab, rabbit anti-enterotoxin antibody; [Anti-Ab]–Enz, peroxidase-labelled anti-rabbit IgG antibody.

titrated by reaction with the enterotoxin-coupled Magnogel and further reaction with a second antibody coupled to an enzyme (Anti-Ab–Enz; anti-rabbit IgG-peroxidase). The enzyme activity of the Magnogel complex is inversely proportional to the initial concentration of toxin present in the sample.

The rabbit IgG antibodies used in MEIA were purified from staphylococcal enterotoxin B antiserum by affinity chromatography using Protein A–sepharose CL-4B gel (Fig. 7). Immunological activity of the purified IgG was then confirmed by Ouchterlony double immunodiffusion in agarose gel (Fig. 8).

Enterotoxin was immobilised to Magnogel and the coupling efficiency of Magnogel was determined by quantifying by the capillary tube assay technique (Fung & Wagner, 1971) the residual unbound enterotoxin B remaining in solution after coupling the toxin to Magnogel.

A typical result of the assay showed an inverse relationship between enzyme activity of the Magnogel complex and the standard toxin concentration over a ten-fold range (Fig. 9; Patel & Gibbs, 1983). The sensitivity of detection of enterotoxin B by the inhibition MEIA is approximately 200 ng toxin/ml, which is twenty-fold less than that of ELISA or RIA but similar to the conventional double-immunodiffusion technique (Holbrook & Baird-Parker, 1975).

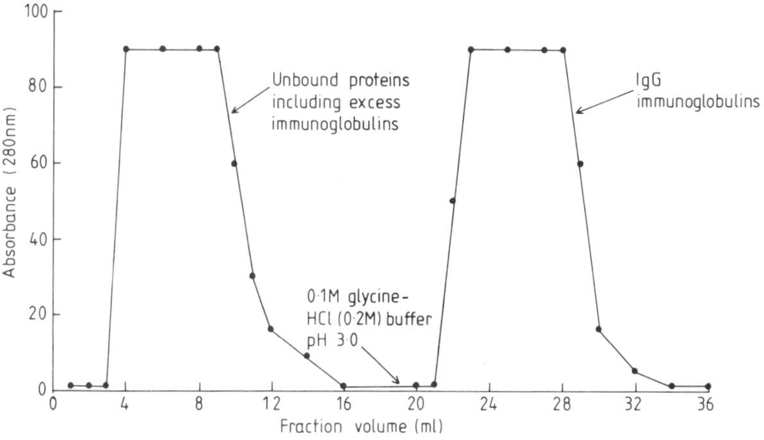

Fig. 7. Purification of rabbit IgG antibodies from enterotoxin B antiserum by affinity chromatography using Protein A–sepharose CL-4B gel. Column conditions: dimensions, 1·5 cm × 1·8 cm; flow rate, 1 ml/min.

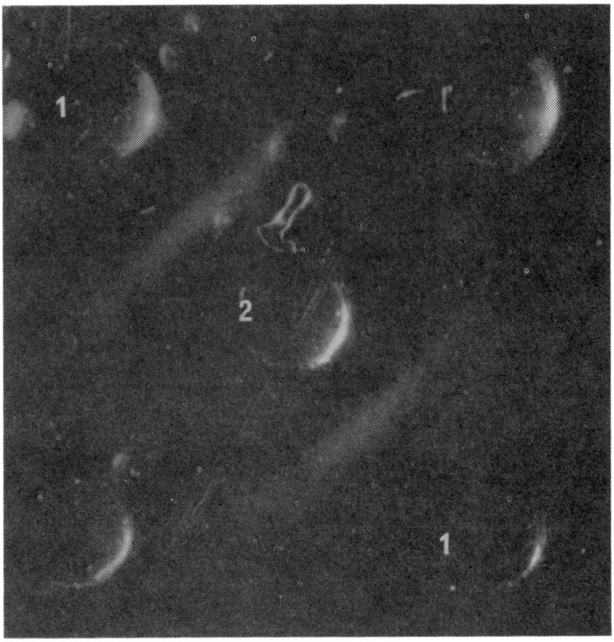

Fig. 8. Double immunodiffusion in agarose gel of purified IgG against 1 μg/ml enterotoxin B: 1, 1:500 dilution of purified IgG; 2, 1 μg/ml enterotoxin B.

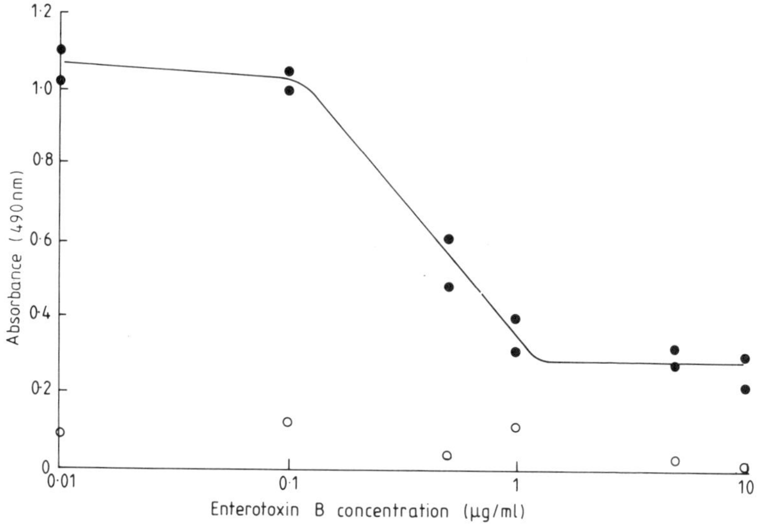

Fig. 9. Typical titration of various concentrations of enterotoxin B against constant level (1:2000 dilution) of rabbit anti-enterotoxin B IgG antibodies. ●, 1:10 dilution of enterotoxin-coupled Magnogel used; ○, 1:10 dilution of enterotoxin-coupled Magnogel, 1:500 dilution of Spicer-Edwards serum versus various concentrations of enterotoxin B (negative serum control).

The detection of enterotoxin B in an enterotoxigenic culture filtrate (prepared according to Donnelly *et al.*, 1967) is shown in Fig. 10. The concentration of enterotoxin B in the culture filtrate estimated by inhibition MEIA was between 50 and 100 μg/ml, whereas that estimated by the double-immunodiffusion technique (Patel, 1982) was approximately 125 μg/ml. Since the inverse relationship between enzyme activity of the Magnogel complex and toxin concentration occurs over a very narrow range (i.e. ten-fold, Fig. 9), inhibition MEIA may be useful for detection rather than quantification of enterotoxin B at concentrations of ⩾0·2 μg toxin/ml. The sensitivity and the dynamic range (i.e. range over which titration between enterotoxin and the specific antibody occurs) may be increased by using a sandwich and possibly both the competitive and non-competitive approaches to MEIA as described for the ELISA techniques. Work is continuing on the application of inhibition MEIA for the estimation of enterotoxin B in foods.

In future, the ELISA techniques (and possibly MEIA) may supersede

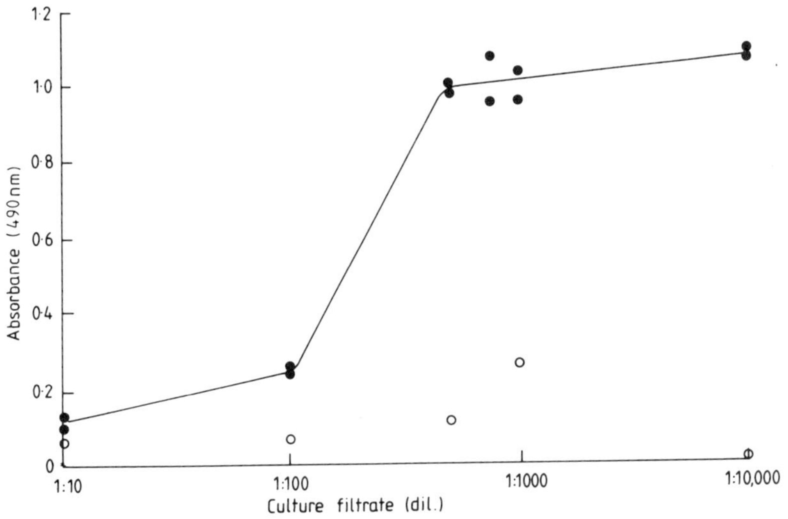

Fig. 10. Titration of various dilutions of enterotoxigenic culture filtrate against constant level (1:2000) dilution of anti-enterotoxin B IgG antibodies: ●, 1:10 dilution of enterotoxin-coupled Magnogel used; ○, 1:10 dilution of enterotoxin-coupled Magnogel, no anti-enterotoxin B antibodies versus various dilutions of culture filtrate (control).

RIA and other serological techniques for estimating not only staphylococcal enterotoxins, but also other microbial toxins and even pathogenic organisms in food, e.g. salmonellae.

Currently a number of laboratories are in the process of producing monoclonal antibodies against staphylococcal enterotoxins (Hubbard & Wiseman, 1983; Meyer *et al.*, 1983). Since monoclonal antibodies are highly specific reagents and will eventually be available cheaply in large quantities, their use, together with the sensitive ELISA techniques, would provide a powerful technique for estimating staphylococcal enterotoxins. The use of monoclonal antibodies could reduce the number of washing stages in both the direct and indirect sandwich types of ELISA technique. For instance, a mixture of antigen with at least two antigenic determinants and a conjugate of monoclonal antibody enzyme could be incubated with a second monoclonal antibody to another determinant adsorbed to a solid phase. Since the antigen has separate binding sites for both the antibodies, it will bind to antibody adsorbed to a solid phase and to the antibody–enzyme conjugate. This avoids competition of both antibodies for the same binding site on the antigen. This type of assay has been reported for the estimation of α-fetoprotein (Ruoslahti & Engvall, 1982).

In recent years the sensitivity of ELISA techniques for estimating macromolecules has been increased in at least three ways:

(1) The enzyme activity of a conjugate is measured as light output in a chemiluminescence reaction (Velan & Halmann, 1978; Pronovost *et al.*, 1982; Whitehead *et al.*, 1983) instead of a colorimetric estimation of chromogenic product.
(2) Utilising better enzymes with a higher turnover number for substrate molecules and a sharper end-point detection, e.g. urease. Urease conjugated antibodies are now available commercially (Sera-Lab Ltd, Sussex).
(3) Amplification of ELISA techniques by utilising biotin/avidin-coupled enzyme systems (Kendall *et al.*, 1983; Yolken *et al.*, 1983). The sensitivity of these techniques is greater than or equal to that of RIA and fluorescence immunoassay techniques.

ACKNOWLEDGEMENTS

This work was supported by the Ministry of Agriculture, Fisheries and Food and is Crown Copyright.

I thank the editors of *Journal of Food Protection* and *Applied Environmental Microbiology* for granting permission to reproduce Figs 2 and 4, respectively.

The advice and assistance of Dr Paul A. Gibbs in the preparation of the manuscript is gratefully acknowledged.

REFERENCES

Ackermann, J. I. & Chesbro, W. (1981) Detection of *Staphylococcus aureus* products in foods using enzyme linked immunosorbent assay and spectrophotometric thermonuclease assay. *Journal of Food Safety*, **3**, 15–25.

Antoine, J. C., Ternynck, T., Rodrigot, M. & Avrameas, S. (1978) Lymphoid cell fractionation on magnetic polyacrylamide-agarose beads. *Immunochemistry*, **15**, 443–52.

Areson, P. D. W., Charm, S. E. & Wong, B. L. (1980) Determination of staphylococcal enterotoxins A and B in various food extracts, using staphylococcal cells containing protein A. *Journal of Food Science*, **45**, 400–1.

BERDAL, B. P. OLSVIK, Ø. & OMLAND, T. (1981) A sandwich ELISA method for detection of *Staphylococcus aureus* enterotoxins. *Acta pathologica microbiologica Scandinavica, Sect. B*, **89**, 411–15.

BERGDOLL, M. S. & REISER, R. F. (1980) Application of radioimmunoassay for detection of staphylococcal enterotoxins in foods. *Journal of Food Protection*, **43**, 68–72.

CROWLE, A. J. (1958) A simplified micro double-diffusion agar precipitin technique. *Journal of Laboratory & Clinical Medicine*, **52**, 784.

DE JONG, P. J. (1983) Simple method for preparation of antibody–enzyme conjugates for enzyme-linked immunosorbent assays. *Journal of Clinical Microbiology*, **17**, 928–30.

DONNELLY, C. B., LESLIE, J. E., BLACK, L. A. & LEWIS, K. H. (1967) Serological identification of enterotoxigenic staphylococci from cheese. *Applied Microbiology*, **15**, 1382–7.

DUBOIS-DALCQ, M., MCFARLAND, H. & MCFARLIN, D. (1977) Protein A-peroxidase: a valuable tool for localization of antigens. *Journal of Histochemistry & Cytochemistry*, **25**, 1201–6.

ENDRESEN, C. (1978) Protein A reactivity of whole rabbit IgG and of fragments of rabbit IgG. *Acta pathologica microbiologica Scandinavica, Sect. C*, **86**, 211–4.

ENGVALL, E. & PERLMANN, P. (1971) Enzyme-linked immunosorbent assay (ELISA). Quantitative assay of immunoglobulin G. *Immunochemistry*, **8**, 871–84.

FEY, H., STIFFLER-ROSENBERG, G., WARTENWEILER-BURKHARD, G., MÜLLER, CHR. & RÜEGG, O. (1982) Der Nachweis von Staphylokokken-Enterotoxinen (SET). *Schweizer Archiv für Tierheilkunde*, **124**, 297–306.

FREED, R. C., EVENSON, M. L., REISER, R. F. & BERGDOLL, M. S. (1982) Enzyme-linked immunosorbent assay for detection of staphylococcal enterotoxins in foods. *Applied & Environmental Microbiology*, **44**, 1349–55.

FUNG, D. Y. C. & WAGNER, J. (1971) Capillary tube assay for staphylococcal enterotoxins A, B and C. *Applied Microbiology*, **21**, 559–61.

GODING, J. W. (1978) Use of staphylococcal protein A as an immunological reagent. *Journal of Immunological Methods*, **20**, 241–53.

GUESDON, J. L. & AVRAMEAS, S. (1977) Magnetic solid phase enzyme-immunoassay. *Immunochemistry*, **14**, 443–7.

GUESDON, J. L., COURCON, J. & AVRAMEAS, S. (1978*a*) Magnetically responsive polyacrylamide agarose beads for the preparation of immunoadsorbents. *Journal of Immunological Methods*, **21**, 59–63.

GUESDON, J. L., THIERRY, R. & AVRAMEAS, S. (1978*b*) Magnetic enzyme immunoassay for measuring human IgE. *Journal of Allergy & Clinical Immunology*, **61**, 23–7.

HOLBROOK, R. & BAIRD-PARKER, A. C. (1975) Serological methods for the assay of staphylococcal enterotoxins. In: *Some methods for microbiological assay*, Board, R. G. & Lovelock, D. W. (eds), SAB Tech. Ser. 8: Academic Press, London, 108–26.

HUBBARD, R. & WISEMAN, A. (1983) Enzyme immunoassay and the use of monoclonal antibodies. *Trends in Analytical Chemistry*, **2**, VII–IX.

KAUFFMAN, P. E. (1980) Enzyme immunoassay for staphylococcal enterotoxin A. *Journal of the Association of Official Analytical Chemists*, **63**, 1138–43.

KENDALL, C., IONESCU-MATIU, I. & DREESMAN, G. R. (1983) Utilization of the biotin/avidin system to amplify the sensitivity of the enzyme-linked immunosorbent assay (ELISA). *Journal of Immunological Methods*, **56**, 329–39.

KOPER, J. W., HAGENAARS, A. M. & NOTERMANS, S. (1980). Prevention of cross-reactions in the enzyme linked immunosorbent assay (ELISA) for the detection of *Staphylococcus aureus* enterotoxin type B in culture filtrates and foods. *Journal of Food Safety*, **2**, 35–45.

KUO, J. K. S. & SILVERMAN, G. J. (1980). Application of enzyme-linked immunosorbent assay for detection of staphylococcal enterotoxins in food. *Journal of Food Protection*, **43**, 404–7.

LENZ, W., THELEN, R., PICKENHAHN, P. & BRANDIS, H. (1983) Detection of enterotoxin in cultures of *Staphylococcus aureus* by the enzyme linked immunosorbent assay (ELISA) and the microslide immunodiffusion. *Zentralblatt für Bakteriologie Parasitenkunde Infektion Hygiene, Abt. I, Originale A*, **253**, 466–75.

LINDMARK, R. (1982) Fixed protein A-containing staphylococci as solid-phase immunoadsorbents. *Journal of Immunological Methods*, **52**, 195–203.

MAGGIO, E. T. (1981) *Enzyme-immunoassay*, CRC Press, Inc, Boca Raton, Florida.

MEYER, R. F., MILLER, L. & MACMILLAN, J. D. (1983) Monoclonal antibody to a determinant common to five serotypes of staphylococcal enterotoxin. *The Society for General Microbiology Quarterly*, **10**, M10.

MILLER, B. A., REISER, R. F. & BERGDOLL, M. S. (1978) Detection of staphylococcal enterotoxins A, B, C, D and E in foods by radioimmunoassay, using staphylococcal cells containing protein A as immunosorbent. *Applied & Environmental Microbiology*, **36**, 421–6.

MINOR, T. E. & MARTH, E. H. (1976) *Staphylococci and their Significance in Foods*, Ch. 7, Elsevier Scientific Publishing Company, New York.

MORITA, T. N. & WOODBURN, M. J. (1978) Homogeneous enzyme-immune assay for staphylococcal enterotoxin B. *Infection and Immunity*, **21**, 666–8.

NISKANEN, A. (1977) *Staphylococcal enterotoxins and food poisoning. Production, properties and detection of enterotoxins.* Technical Research Centre of Finland, *Publication No. 19*, Valtion Teknillinen Tutkimuskeskus, Espoo, p. 50.

NOTERMANS, S., VERJANS, H. L., BOL, J. & VON SCHOTHORST, M. (1978) Enzyme linked immunosorbent assay (ELISA) for determination of *Staphylococcus aureus* enterotoxin type B. *Health Laboratory Science*, **15**, 28–31.

OLSVIK, Ø., MYHRE, S., BERDAL, B. P. & FOSSUM, K. (1982) Detection of staphylococcal enterotoxin A, B and C in milk by an ELISA procedure. *Acta Veterinaria Scandinavica*, **23**, 204–10.

O'SULLIVAN, M. J., BRIDGES, J. W. & MARKS, V. (1979) Enzyme immunoassay: a review. *Annals of Clinical Biochemistry*, **16**, 221–39.

PATEL, P. D. (1982) Evaluation of a homogeneous enzyme-linked immunoassay for staphylococcal enterotoxin B. *Chemistry & Industry*, 18 Dec., 979–81.

PATEL, P. D. & GIBBS, P. A. (1983) Development of an inhibition magnetic enzyme immunoassay technique (IMEIAT) for estimation of staphylococcal enterotoxin B. *Leatherhead Food Research Association Technical Notes*, No. 6, p. 5.

PATEL, P. D., WOOD, J. M. & GIBBS, P. A. (1983) Development of a magnetic enzyme immunoassay technique (MEIAT) for estimation of staphylococcal enterotoxin B IgG-type antibodies. *Leatherhead Food Research Association Technical Notes*, No. 6, p. 1.

PRONOVOST, A. D., BAUMGARTEN, A. & ANDIMAN, W. A. (1982) Chemiluminescent immunoenzymatic assay for rapid diagnosis of viral infections. *Journal of Clinical Microbiology*, **16**, 345–9.

REISER, R., CONAWAY, D. & BERGDOLL, M. S. (1974) Detection of staphylococcal enterotoxin in foods. *Applied Microbiology*, **27**, 83–5.

RUOSLAHTI, E. & ENGVALL, E. (1982) Monoclonal antibodies in immunoassays. *Clinical Immunology Newsletter*, **3**, 139–42.

SAUNDERS, G. C. & BARTLETT, M. L. (1977) Double-antibody solid-phase enzyme immunoassay for the detection of staphylococcal enterotoxin A. *Applied & Environmental Microbiology*, **34**, 518–22.

SIMON, E. & TERPLAN, G. (1977) Nachweis von Staphylokokken Enterotoxin B mittles ELISA-Test. *Zentralblatt für Veterinar Medizin, Reihe B*, **24**, 842–4.

STIFFLER-ROSENBERG, G. & FEY, H. (1978) Simple assay for staphylococcal enterotoxins A, B and C: modification of enzyme-linked immunosorbent assay. *Journal of Clinical Microbiology*, **8**, 473–9.

VAN DER OUDERAA, F. & HAAS, H. (1981) Use of immunoassays to detect enterotoxin B of *Staphylococcus aureus* in foods. *Antonie van Leeuwenhoek*, **47**, 186–7.

VELAN, B. & HALMANN, M. (1978) Chemiluminescence immunoassay; a new sensitive method for determination of antigens. *Immunochemistry*, **15**, 331–3.

VOLLER, A. & BIDWELL, D. (1980) *The Enzyme Linked Immunosorbent Assay* (*ELISA*), Vol. 2, MicroSystems Ltd, Summerfield House, Vale, Guernsey, UK.

VOLLER, A., BIDWELL, D. E. & BARTLETT, A. (1979) *The Enzyme Linked Immunosorbent Assay* (*ELISA*), Dynatech Europe, Borough House, Rue du Pré, Guernsey, UK.

WEEMEN, B. K. V. & SCHUURS, A. H. W. M. (1971) Immunoassay using antigen–enzyme conjugates. *FEBS Letters*, **15**, 232–6.

WHITEHEAD, T. P., THORPE, G. H. G., CARTER, T. J. N., GROUCUTT, C. & KRICKA, L. J. (1983) Enhanced luminescence procedure for sensitive determination of peroxidase-labelled conjugates in immunoassay. *Nature*, **305**, 158–9.

WISDOM, E. B. (1976) Enzyme-immunoassay. *Clinical Chemistry*, **22**, 1243–55.

YOLKEN, R. H., LEISTER, F. J., WHITCOMB, L. S. & SANTOSHAM, M. (1983) Enzyme immunoassays for the detection of bacterial antigens utilizing biotin-labelled antibody and peroxidase biotin-avidin complex. *Journal of Immunological Methods*, **56**, 319–27.

SESSION III

Application to Small Molecules

12
An ELISA for the Analysis of the Mycotoxin Ochratoxin A in Food

M. R. A. MORGAN, R. MCNERNEY and H. W.-S. CHAN

AFRC Food Research Institute, Norwich, UK

INTRODUCTION

The ochratoxins are toxic secondary metabolite products of a number of fungal species in the *Aspergillus* and *Penicillium* genera. These fungi are found as natural contaminants of a number of agricultural commodities, but only under certain conditions, such as inadequate storage, will toxin synthesis occur. In temperate climates the *Penicillium* group will normally be the most important toxin source. Ochratoxins have been identified in a wide range of food and food materials, such as corn, oats, wheat, barley, peanuts and other nut products, and mixed grain feeds. Animals consuming contaminated feed may concentrate toxin in certain tissues, providing further possibilities for human exposure. Ochratoxin A, the major toxin of the group, is a 7-carboxy-5-chloro-8-hydroxy-3,4-dehydro-3-R-methyl isocoumarin amide of L-β-phenylalanine (Fig. 1). The methyl and ethyl esters of ochratoxin A (the latter known as ochratoxin C) appear to show similar toxicity to that of the parent compound, though their importance as natural fungal products outside laboratory cultures remains to be established. Ochratoxin B, the dechloro analogue, is much less toxic (Steyn & Holzapfel, 1967).

Ochratoxin A is a potent nephrotoxin. It has been linked closely with mycotoxic porcine nephropathy (Krogh, 1978), the most common symptoms of which are polydypsia and polyuria. Interstitial fibrosis and tubular atrophy can be observed in the kidney at slaughter. The toxin is also believed to be involved in the frequently fatal human disease Endemic Balkan Nephropathy found in parts of Bulgaria, Roumania and Yugoslavia (Krogh *et al.*, 1977; Pavlovic *et al.*, 1979). Other toxicological

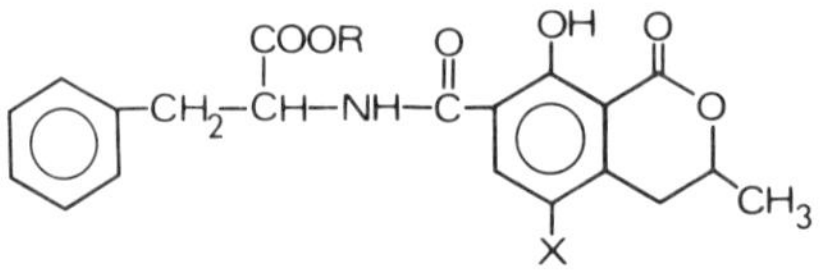

	R	X
Ochratoxin A	H	Cl
Ochratoxin B	H	H
Ochratoxin C	CH_2CH_3	Cl

Fig. 1. The structures of ochratoxins A, B and C.

responses that have been reported include effects on the liver (Purchase & Theron, 1968), bone abnormalities in poultry (Huff *et al.*, 1980), intestinal problems in poultry, mice and rats (Huff *et al.*, 1974; Galtier *et al.*, 1974; Galtier *et al.*, 1975), and immunosuppression at low doses in mice (Creppy *et al.*, 1982). Teratogenic actions have been reported in the mouse (Hood *et al.*, 1978), rat (Still *et al.*, 1971), hamster (Hood *et al.*, 1976), and chick embryo (Giliani *et al.*, 1978).

Many animals appear to metabolise ochratoxin A by hydrolysis to produce ochratoxin α and L-phenylalanine (Nel & Purchase, 1968). To a much lesser extent, hydroxylation to 4- and 10-hydroxyochratoxin derivatives may also occur (Syvertsen & Størmer, 1983). The metabolites are non-toxic. Sixty-six percent of an oral dose of ochratoxin A was absorbed by pigs (Galtier *et al.*, 1981), with concentration mainly in the kidney. Significant residues were also found in liver and muscle. The long half-life in pigs (89 h) indicates the likely long-term persistence of the toxin.

ANALYSIS OF OCHRATOXIN A — BIOASSAYS AND CHEMICAL METHODS

Various bioassays for mycotoxins have been described, such as the use of brine shrimp larvae (Harwig & Scott, 1971) and day-old ducklings (Steyn

& Holzapfel, 1967). In both cases death is the response observed. Isolated cells rather than whole organisms can be used for a similar purpose — cytomorphological changes caused by mycotoxins have been observed in cell cultures of various rat tissues (Umeda, 1971). More recently much interest has been shown in the possibilities of using bacterial bioluminescence as an indicator of mycotoxin presence (e.g. Yates & Porter, 1982). These are potentially simple and inexpensive procedures, but share the disadvantages of all bioassays. Their lack of specificity and sensitivity, the large variability inherent in bio-responses, and the frequent occurrence of false-positive results given by food extracts have all restricted wider use. The major advantage of these assays, one often overlooked, is that they actually measure toxicity in some form.

Thin-layer chromatography (Nesheim *et al.*, 1973) and high performance liquid chromatography (Hunt *et al.*, 1978) have both been widely exploited for ochratoxin A determination in food. One of the advantages of both these methods is that it is possible, under the appropriate conditions, to quantitate several toxins at the same time (Roberts & Patterson, 1975). Sample pre-purification before chromatography is normally tedious and time-consuming, a major limitation when it is required to assay large sample numbers. False-positive results occur, and this means that positive results need to be confirmed by re-analysis using a different analytical procedure.

IMMUNOASSAY OF OCHRATOXIN A

Three laboratories have described immunoassays for ochratoxin A. The first report (Aalund *et al.*, 1975) was a preliminary investigation of a radioimmunoassay for the toxin, using an ^{125}I-labelled egg albumen–ochratoxin A conjugate as the labelled antigen. The anti-ochratoxin A antiserum was raised in rabbits against an ochratoxin A–bovine IgG conjugate, synthesised by a carbodiimide method utilising the carboxylic acid function of the toxin. The antiserum produced was used at a dilution of 1 in 200 (v/v) in generating a standard curve with a limit of detection of 20 ng ochratoxin A. Two overnight incubation steps were incorporated in the assay, the second of which accomplished the separation of antibody-bound and free fractions by precipitating the primary antibody with a second antibody, directed against rabbit IgG.

Antisera against ochratoxin A were produced and characterised by Chu *et al.* (1976). Several potential immunogens were synthesised using a carbodiimide as coupling agent, the carrier proteins involved being two polymers of poly-lysine, lysozyme and bovine serum albumin. The albumin conjugate produced usable antisera after injection into rabbits, giving a radioimmunoassay standard curve with a limit of detection of 2·5 ng. ^{3}H-ochratoxin A, which is not available commercially, was used as label. This antiserum was also used in a microtitre plate enzyme-linked immunosorbent assay (ELISA) (Pestka *el al.*, 1981). It was immobilised to the surface of the microtitre plate wells at a dilution of 1 in 200 (v/v). Competition for the binding sites takes place between the ochratoxin A of the sample and the ochratoxin A to which the enzyme has been covalently bound. Separation of antibody-bound and free fractions is a simple plate washing procedure, and detection of antibody-bound label requires addition of substrate. The limit of detection for the standard curve was 25 pg.

The third reported immunoassay (Morgan *et al.*, 1982; Morgan *et al.*, 1983), which is also a microtitre plate ELISA, is the only one to describe the determination of ochratoxin A in biological material. The avoidance of radiolabels, by using non-isotopic methods, is of general importance to the food analyst. The attraction of the microtitre plate format is that comparatively cheap, robust assays can be produced that can be used at many levels, ranging from full automation in the well-equipped laboratory to application in the field with visual assessment of results. The particular type of ELISA we have used is a double antibody method, an approach hitherto used relatively infrequently for non-immunogenic small molecules. More commonly, a direct assay has been employed where enzyme-labelled and free hapten (sample or standard) compete for antibody immobilised on the surface of the wells of the microtitre plate. We prefer to immobilise the hapten and incubate with primary antisera and free hapten in the sample or standard. Antibodies subsequently bound are detected with an enzyme-labelled second antibody. There are several advantages to this procedure:

(i) The direct method may be inefficient in its use of primary antisera. This is because antibody immobilisation might not be quantitative and because antibodies could be bound in such a manner as to be unavailable for hapten binding activity.

(ii) The direct method requires the synthesis of active enzyme-labelled hapten, perhaps in a particular molar ratio for maximum assay sensitivity. This presents unnecessary problems that can be avoided in the indirect method. Here the synthesis of the protein–hapten conjugate used to form the immobilised phase is relatively undemanding — purification is not essential and alteration of protein structure during reaction is unimportant.

(iii) The direct method requires incubation of sample and enzyme label together. In food analysis, the sample could contain enzyme inhibitors or promotors which might cause sample interference. Sample and enzyme appear in separate incubation stages in the indirect ELISA.

(iv) There are potential advantages in both the wide working range and theoretically superior limit of detection of the double antibody method.

The procedure for the indirect ELISA is as follows:

(1) Take microtitre plate pre-coated with ochratoxin A.
(2) *Add sample or standard and rabbit anti-ochratoxin A antiserum* to appropriate wells. Incubate. Wash.
(3) *Add enzyme-labelled anti-rabbit IgG antiserum* to each well. Incubate. Wash.
(4) *Add enzyme substrate* to each well. Incubate. If required, stop reaction.
(5) Measure optical densities of individual wells.

Ochratoxin A is immobilised ('coated') onto the plate wells by non-covalent adsorption of a protein–toxin conjugate. In this case, keyhole limpet haemocyanin was used as the protein carrier and coupling was effected using a carbodiimide reagent. The particular protein used as carrier for coating the wells with antigen is unimportant as long as no interaction occurs between it and the primary antisera, the food extract or the labelled second antisera.

Our anti-ochratoxin A antiserum was raised in rabbits to a conjugate of ochratoxin A with bovine serum albumin, synthesised by the mixed anhydride method. It is used in the ELISA at a dilution of 1 in 25 000 (v/v). High dilutions such as this are important for the elimination of non-specific effects and for continuity. It also enables wider dissemination of assays.

IMMUNOASSAY OF OCHRATOXIN A IN FOOD

Our anti-ochratoxin A antiserum is extremely specific, giving very low cross-reaction figures for a variety of metabolites and structurally-related compounds (see Table 1). This specificity means that the antiserum is

Table 1
Cross-reactions of Various Compounds in the ELISA System

Compound	*Cross-reaction* (%)
Ochratoxin A	100
Ochratoxin B	0·5
Ochratoxin α	2·4
Coumarin	<0·000 05
4-Hydroxycoumarin	<0·000 05
Phenylalanine	<0·000 05
(4R)-4-Hydroxyochratoxin A	1·3
10-Hydroxyochratoxin A	1·4

able to pick out ochratoxin A selectively, even in a crude mixture. Sample preparation prior to assay should, therefore, be minimal. That this turns out to be the case in practice is seen in the assay for ochratoxin A in barley (Morgan *et al.*, 1983). Ground barley (5 g) is shaken with 0·1 M-phosphoric acid (5 ml) and chloroform (60 ml). An aliquot (10 ml) of the extract is removed, taken to dryness and re-dissolved in the assay buffer. This solution can then be assayed directly in the ELISA. The procedure is a simple one, consisting only of the solubilisation of the toxin. When alternative methods of analysis are employed, the solubilisation is only the first of several stages required before purity is sufficient to allow quantification.

The simplicity of the extraction required for the barley ELISA is advantageous in two ways. Firstly, and obviously, samples can be extracted and analysed at a quicker rate. Speed is an important factor in mycotoxin analysis where surveillance and quality control programmes, using necessarily rigid sampling schemes, generate large sample numbers. Secondly, a reduction in the number of manipulations will have a beneficial effect on reproducibility and other assay characteristics. Thus recovery of toxin added to barley is virtually quantitative. Manipulation of amounts of contaminated barley extracted, the volumes of extract assayed and dilutions of contaminated barley with uncontaminated material all yield a linear response for toxin analysed. The sensitivity of the assay is potentially as low as 60 pg/g barley. The assay has been tested in several trials alongside both thin-layer and high performance liquid chromatographic methods. Correlation of results has been impressive.

Preliminary results suggest that the immunoassay can be extended to the analysis of ochratoxin A in porcine kidney. Conventional methods are extremely laborious, and a large proportion of samples are contaminated only at very low levels or at levels below assay detection limits. Our results suggest that with only a small alteration in the methodology used in the barley extraction procedure, kidney samples can also be analysed by ELISA (R. McNerney & M. R. A. Morgan, unpublished results). The assay is rapid, capable of extracting and analysing at least 100 samples per week. Again, recovery of added toxin is quantitative. Ochratoxin A-contaminated samples show linearity of response with dilution (both with assay buffer and with kidney samples containing no detectable toxin).

The application of the ochratoxin A ELISA clearly illustrates the potential of the immunological analytical approach in the field of mycotoxin analysis. It is envisaged that a range of immunoassays for these toxins in particular food materials will shortly be available to the analyst.

REFERENCES

Aalund, O., Brunfeldt, K., Hald, B., Krogh, P. & Poulsen, K. (1975) A radioimmunoassay for ochratoxin A: a preliminary investigation. *Acta Pathologica Microbiologica Scandinavica, Section C*, **83**, 390–2.

Chu, F. S., Chang, F. C. C. & Hindshill, R. D. (1976) Production of antibody against ochratoxin A. *Applied and Environmental Microbiology*, **31**, 831–5.

Creppy, E. E., Lorkowski, G., Roschenthaler, R. & Dirheimer, G. (1982) Kinetics of the immunosuppressive action of ochratoxin A on mice. *Proceedings of the V International IUPAC Symposium on Mycotoxins and Phycotoxins*, Vienna, 289–92.

Galtier, P., More, J. & Bodin, G. (1974) Toxines d'*Aspergillus ochraceous* Wilhelm. III. Toxicité aigue de l'ochratoxine A chez le rat et la souris adultes. *Annales de Recherches veterinaires*, **5**, 233–47.

Galtier, P., Bodin, G. & More, J. (1975) Toxines d'*Aspergillus ochraceous* Wilhelm. IV. Toxicité de l'ochratoxine A par administration orale prolongée chez le rat. *Annales de Recherches veterinaires*, **6**, 207–18.

Galtier, P., Alvinerie, M. & Charpenteau, J. L. (1981) The pharmacokinetic profiles of ochratoxin A in pigs, rabbits and chickens. *Food and Cosmetics Toxicology*, **19**, 735–8.

Giliani, S. H., Bancroft, J. & Reily, M. (1978) Teratogenicity of Ochratoxin A in chick embryos. *Toxicology and Applied Pharmacology*, **46**, 543–6.

Harwig, J. & Scott, P. M. (1971) Brine shrimp (*Artemia salina* L.) larvae as a screening system for fungal toxins. *Applied Microbiology*, **21**, 1011–16.

Hood, R. D., Naughton, M. J. & Hayes, A. W. (1976) Prenatal effects of ochratoxin A in hamsters. *Teratology*, **13**, 11–14.

HOOD, R. D., KUCNZUK, M. H. & SZCYECH, G. M. (1978) Effects in mice of simultaneous prenatal exposure to ochratoxin A and T-2 toxin. *Teratology*, **17**, 25–30.

HUFF, W. E., WYATT, R. D., TUCKER, T. L. & HAMILTON, P. B. (1974) Ochratoxicosis in the broiler chicken. *Poultry Science*, **53**, 1585–91.

HUFF, W. E., DOERR, J. A., HAMILTON, P. B., HAMANN, D. D., PETERSON, R. E. & CIEGLER, A. (1980) Evaluation of bone strength during aflatoxicosis and ochratoxicosis. *Applied and Environmental Microbiology*, **40**, 102–7.

HUNT, D. C., BOURDON, A. T., WILD, P. J. & CROSBY, N. T. (1978) Use of high performance liquid chromatography combined with fluorescence detection for the identification and estimation of aflatoxins and ochratoxin in food. *Journal of the Science of Food and Agriculture*, **29**, 234–8.

KROGH, P. (1978). Causal associations of mycotoxic nephropathy. Thesis. *Acta Pathologica Microbiologica Scandinavica, Section A*, Supplement 269, 1–28.

KROGH, P., HALD, B., PLESTINA, R. & CEOVIC, S. (1977) Balkan (endemic) nephropathy and foodborn ochratoxin A: preliminary results of a survey of foodstuffs. *Acta Pathologica Microbiologica Scandinavica, Section B*, **85,** 238–40.

MORGAN, M. R. A., MATTHEW, J. A., MCNERNEY, R. & CHAN, H. W.-S. (1982) The immunoassay of ochratoxin A. *Proceedings of the V International IUPAC Symposium on Mycotoxins and Phycotoxins*, Vienna, 32–5.

MORGAN, M. R. A., MCNERNEY, R. & CHAN, H. W.-S. (1983) Enzyme-linked immunosorbent assay of ochratoxin A in barley. *Journal of the Association of Official Analytical Chemists*, **66**, 1481–4.

NEL, W. & PURCHASE, I. F. H. (1968) The fate of ochratoxin A in rats. *Journal of the South African Chemical Institute*, **21**, 87–8.

NESHEIM, S., HARDIN, N. F., FRANCIS, O. J. & LANGHAM, W. S. (1973) Analysis of ochratoxins A and B and their esters in barley, using partition and thin layer chromatography. I. Development of the method. *Journal of the Association of Official Analytical Chemists*, **56**, 817–21.

PAVLOVIC, M., PLESTINA, R. & KROGH, P. (1979) Ochratoxin A contamination of foodstuffs in an area with Balkan (endemic) Nephropathy. *Acta Pathologica Microbiologica Scandinavica, Section B*, **87**, 243–6.

PESTKA, J. J., STEINERT, B. W. & CHU, F. S. (1981) Enzyme-linked immunosorbent assay for detection of ochratoxin A. *Applied and Environmental Microbiology*, **41**, 1472–4.

PURCHASE, I. F. H. & THERON, J. J. (1968) The acute toxicity of ochratoxin A to rats. *Food and Cosmetic Toxicology*, **6**, 479–83.

ROBERTS, B. A. & PATTERSON, D. S. P. (1975) Detection of twelve mycotoxins in mixed animal feedstuffs using a novel membrane cleanup procedure. *Journal of the Association of Official Analytical Chemists*, **58**, 1178–81.

STEYN, P. S. & HOLZAPFEL, C. W. (1967) The isolation of the methyl and ethyl esters of ochratoxin A and B metabolites of *Aspergillus ochraceous*. *Journal of the South African Chemical Institute*, **20**, 186–9.

STILL, P. E., MACKLIN, A. W., RIBELIN, W. E. & SMALLEY, E. B. (1971) Relationship of ochratoxin A to foetal death in laboratory and domestic animals. *Nature*, **234**, 563–4.

SYVERTSEN, C. & STØRMER, F. C. (1983) Oxidation of two hydroxylated ochratoxin A metabolites by alcohol dehydrogenase. *Applied and Environmental Microbiology*, **45**, 1701–3.

UMEDA, M. (1971) Cytomorphological changes of cultured cells from rat liver, kidney and lung induced by several mycotoxins. *Japanese Journal of Experimental Medicine*, **4**, 195–207.

YATES, I. E. & PORTER, J. K. (1982) Bacterial bioluminescence as a bioassay for mycotoxins. *Applied and Environmental Microbiology*, **44**, 1072–5.

13

The Use of Immunoassay for Monitoring Anabolic Hormones in Meat

M. J. WARWICK, M. L. BATES and G. SHEARER

Ministry of Agriculture, Fisheries and Food, Food Science Laboratory, Norwich, UK

INTRODUCTION

The claimed advantages of immunoassay are sensitivity, specificity, simplicity, cheapness and a high sample throughput, and many of these can be realised when dealing with relatively simple and consistent matrices, such as plasma. The usefulness of this technique is now well documented with, for example, therapeutic drug monitoring, where speed of analysis can be crucial (Marks, 1981). However, when analysing food commodities, one is often dealing with complex materials whose composition may be highly variable. This exacerbates problems inherent in immunoassay and this chapter illustrates some of these, as well as describing methods to overcome or accommodate them, by reference to our experiences in monitoring for the presence of anabolic hormones in meat and other animal products.

DETERMINATION OF ANABOLIC HORMONES IN ANIMAL TISSUE

Anabolic hormones are used extensively in the farming industry to increase growth rate and food conversion by meat producing animals. Both oestrogens and androgens are used and, since their effects are hormonal, they are active at very low levels. For maximum efficiency, consistent low levels should be maintained over relatively long periods of time (Lu *et al.*, 1975). At the present time, there is an increasing tendency to use 'natural' hormones such as oestradiol; this is used at such a low dosage that, in treated animals, elevated levels above the normal range cannot easily be distinguished. The synthetic hormones, which are cheap and very potent, can be more readily detected. At least one of these, diethylstilboestrol (DES), has been shown to be a carcinogen (IARC, 1979); because of this, the stilbenes as a group (of which DES is a member) have been banned throughout the EEC (1981*a*). The other two major synthetic hormones are zeranol, another oestrogen, and trenbolone, an androgen.

Figure 1 shows the structures of the commonly used natural and synthetic hormones. If maximum residue levels are fixed, the analytical methods required to detect these residues need to be quantitative. If the

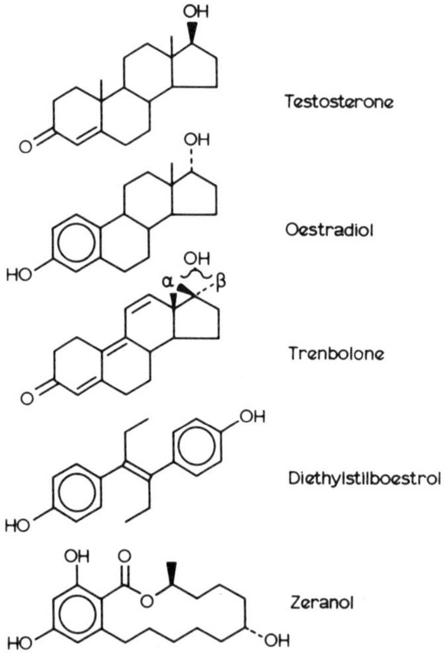

Fig. 1. Anabolic hormones.

use of a compound is no longer allowed, it is merely sufficient to demonstrate the presence of its residue. Table 1 illustrates the levels of stilbenes to be found in various tissues of cattle given 60 mg as an implant. The concentration in excretory products and tissue will be

Table 1
Levels of Stilbenes likely to be Present in Tissues from Animals Slaughtered before Expiry of the Normal Withdrawal Period

Tissue	*Concentration (ng/g or ng/ml)*
Muscle	0–0·2
Liver and kidney	0–1·0
Urine	0·5
Faeces and bile	5·0

related to the time since application, and the concentration in muscle will be dependent on many factors (Harwood *et al.*, 1980; Heitzman & Harwood, 1983). What is apparent is that the low levels to be detected mean that an immunoassay is the only practical approach at present.

The immunoassay that has been most commonly applied to this problem is a type II (i.e. competitive) assay (Ekins, 1981) usually in the form of a radioimmunoassay. As used in the work reported here, the assay involves a fixed amount of tritiated antigen competing with an unknown concentration of antigen in the sample for a limited number of binding sites in the antibody fraction. After incubation, free antigen is removed using dextran coated charcoal and the antibody bound labelled antigen measured in a liquid scintillation counter. The concentration of antigen in the sample is determined by reference to a standard curve performed with each assay.

In the case of steroid hormones, antisera are produced to an immunogen of the antigen conjugated to a carrier protein such as BSA. By judicious choice of linkage point, it is possible to some extent to tailor the cross-reactivity and other properties of the antibody so produced (see Chapter 15). Table 2 shows the specificities of three anti-stilbene antisera produced by Hoechst, the first two of which are commercially available. The first, 6139, was raised against hexoestrol, but shows significant cross-reaction with DES and dienoestrol; it is used in a general screening test for stilbenes. The second antiserum, 254, was raised against DES and has low cross-reactivity with other stilbenes. If a positive result is obtained using antiserum 6139, by subsequent use of 254 and diluted 6139, it is possible to distinguish between the stilbenes.

Anti-trenbolone antisera were kindly supplied by Dr Heitzman's group at the AFRC Institute for Research into Animal Diseases, Compton.

Table 2
Antisera Specificities (% Cross-reactivity)*

	Antiserum code		
	6139	*254*	*6024*
DES	42	100	44
Hexoestrol	100	9	100
Dienoestrol	11	4	13

*Using ^{3}H-DES as the labelled antigen and at dilutions giving equivalent sensitivity to DES.

These workers raised antisera specific to either α- or β-trenbolone, the latter being the administered agent and the former the major metabolite. Tritiated α- and β-trenbolone were supplied by Roussel-Uclaf.

Dr Heitzman's group has also raised several antisera to zeranol which have varying cross-reaction with the parent compound, the mycotoxin zeranolenone, and its metabolite, zeranolenol. The use of monoclonal antibodies has reduced this cross-reaction to very low levels.

The choice of tissue to analyse is determined largely by the information required. Where substances are banned, then it is only necessary to demonstrate significant concentrations of the compounds in any tissue to show that they have been used. However, where maximum residue levels are set, then these are only meaningful when that particular tissue is examined. A more complex situation arises where there is a need to determine exposure data; in this case, all tissues likely to be eaten must be analysed and the concentration of analyte determined in each one.

Whatever the tissue selected, extracts of it will almost always result in non-parallelism. This phenomenon occurs when a component from the tissue changes the extent of binding between antiserum and antigen, either increasing or decreasing it. The result is that binding in the presence of tissue extract, when compared with the standard curve, indicates an apparent high positive or negative result which may vary with analyte concentration.* There may be other more indirect effects of tissue extracts, such as colour affecting counting efficiency, or agents decreasing the efficacy of the charcoal and consequently the non-specific

*It is possible to reduce this phenomenon by diluting the standards in an antigen-free extract of the material being assayed though this is not a practical proposition when undertaking a survey of samples of variable matrix.

binding (radioactivity remaining in solution but not bound to antiserum). Problems arise where there is a direct effect by other molecule(s) on the antibody–antigen reaction, or one which modifies the kinetics or equilibrium of the interaction. If the competitor has a lower affinity for the antiserum than the antigen, as it should have, then dilution of the sample and changes in dilution of the antiserum should improve the result. However, this approach is limited by sensitivity requirements, which, when monitoring for a prohibited compound, should be as high as possible.

TISSUE PURIFICATION PROCEDURES

An alternative approach is to purify the tissue extract and attempt to isolate the analyte from the inhibitors. This is one of the approaches we have used. Figure 2 outlines a typical purification procedure to isolate stilbenes from muscle or organs (Hoffmann, 1978). Essentially, the tissue is homogenised, treated with glucuronidase/sulphatase to hydrolyse conjugates, extracted with a suitable solvent, and subjected to various solvent/solvent partitions to purify the antigens. A similar approach can be used to isolate zeranol, which is also phenolic. Trenbolone requires the same purification procedure, but in this case the chloroform phase is retained after partition with alkali, rather than the alkali phase. The method is clearly protracted and labour intensive; one person can prepare 15 samples per week (each in triplicate) together with appropriate numbers of spiked samples (standard additions) and solvent blanks, the cost being £50 per sample. It is obvious that to monitor a statistically valid proportion of the UK's slaughterhouse output ($\sim$25 million pigs per year, for example) would be an extremely expensive operation and require a large number of staff.

However, the method works well for muscle, kidney and liver; results obtained in our laboratory for kidney are shown in Table 3. First, it should be noted that there is a significant effect from solvent alone. This is one of the problems when extracting and purifying the antigen prior to analysis, since this figure, and more importantly its variance, can affect the variance of the final estimate. A separate solvent blank is carried out with each assay to obviate this. It is important that more problems are not introduced by using certain solvents than are solved. The quality of solvents required depends on the stage in the purification procedure at which they are used. They should always be re-distilled or of at least 'Analar' or equivalent grade, but those used just prior to the im-

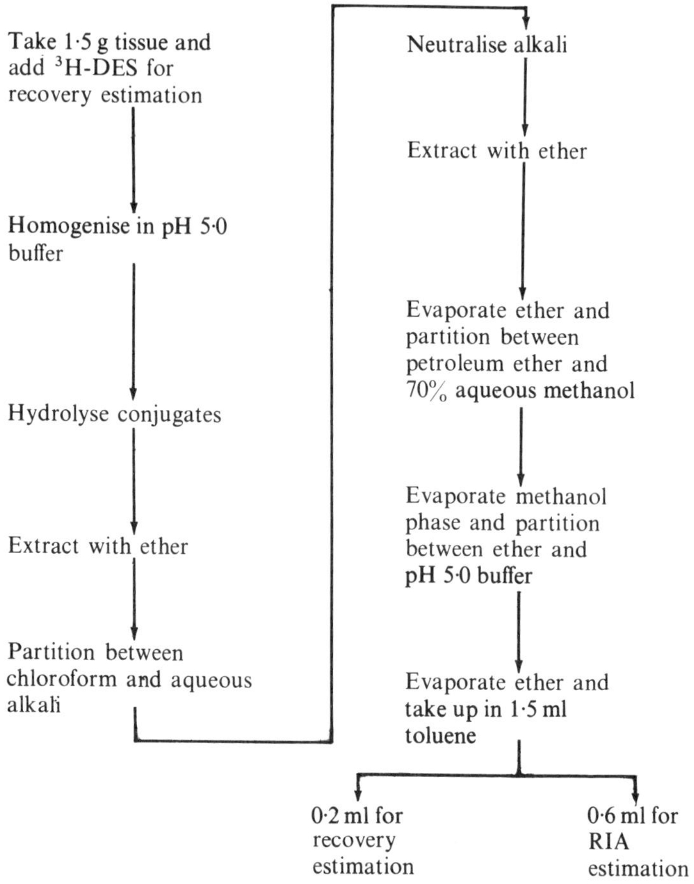

Fig. 2. Procedure for extraction of stilbenes from tissue prior to radioimmunoassay (after Hoffmann, 1978).

munoassay may need to be of 'Aristar' quality or equivalent. The most important data in Table 3 are the values for tissue background and their variation, viz. 30 ± 30 pg/g. This is the apparent content of DES in tissue known not to contain that compound. It is the noise signal of the determination and is the result of inhibition of binding of antigen to antibody by some agent in the tissue extract. The figure for muscle is a little better at 20 ± 20 pg/g (mean $\pm$ SD).

The noise level of the assay is particularly important when attempting

Table 3
Basic Data on Radioimmunoassay of Stilbenes Extracted from Veal Kidney

	Mean	*SD*	*Range*	*n*
Solvent blank value (pg)	16	13	1–68	50
Tissue background value (pg/g)	30	33	−44 to 145	250
Results of spiked samples (pg/g)	265[a]	121	53–576	50
Recovery (%)	66	10	30–92	150

[a]Should be 266 pg/g.

to define which values are significantly positive; this value is the limit of detection of the assay. Decision strategy must be based on scientifically valid methods aimed at minimising the risks of false positives or negatives. If one assumes that the process noise and analytical signals representing true positives have Gaussian distributions, then the detection limit will not only depend upon the means and spread of these parameters, but also on the extent that their distributions overlap. Clearly where noise and positive signals are close, and their distributions overlap, attempting to avoid false positive decisions will result in a major risk of increasing the number of false negative ones. This is true of muscle tissue, where the mean background is 10% of the maximum expected level of contamination. It is then necessary to build into the decision strategy estimates of the relative importance (cost) of false positives and false negatives. Decision strategy is greatly helped if, in the original method development, an attempt is made to ensure that the process noise is as far as possible below the lowest expected positive signal. This requires a good purification procedure. Table 3 also shows the recovery of stilbene through the isolation process; because of the large variation in this value, it is unwise to apply the mean recovery value to all samples. When accurate quantification is required, it is necessary to have a means of determining the recovery for each replicate of a sample. This is best done by spiking the tissue with a low level of radiolabelled analyte and determining the radioactivity at the end of the procedure; account must then be taken of the extra radioactivity entering the radioimmunoassay.

Although the above methods and results are adequate for precise quantification of stilbenes in edible tissue, a cheaper and faster method was required for monitoring purposes; positive samples could then be assayed quantitatively as above. The best tissues to choose for monitoring purposes are obviously those in which the analyte concentrates, i.e.

excretory tissue, and, of these, only bile can be conveniently and hygienically sampled at a slaughterhouse. Figure 3 shows the effect of using accepted procedures to extract bile on the standard curve for stilbenes obtained with antiserum 6139 (EEC, 1981*b*). Experimentally, this was achieved by repeating the standard curve with each standard

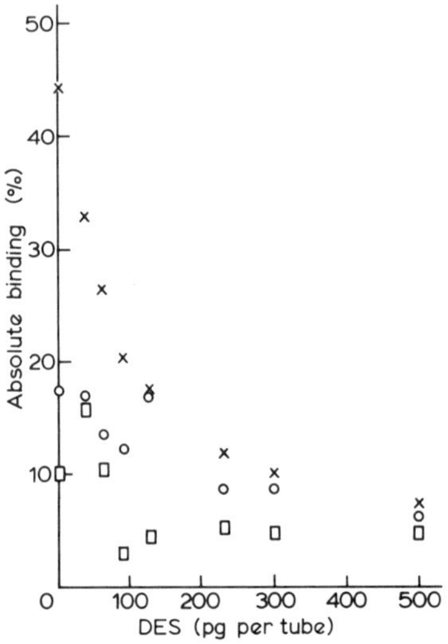

Fig. 3. Standard curves using DES and 6139 antiserum. (i) ×, Normal standard curve; (ii) □, standard curve in presence of 50 μl ether extracted bile; (iii) ○, as in (ii) with subsequent partition between aqueous methanol and petroleum ether. All results corrected for NSB.

containing the equivalent of 50 μl of suitably extracted bile (Hoffmann, 1978). Simple ether extraction resulted in extensive non-parallelism and very variable results and clearly this method of sample preparation is completely unsuitable. A logit/log transformation of the data for the standard curve gave a correlation coefficient of −0·95, and for the curve containing ether extract of bile the result was −0·08. Subsequent liquid/liquid partition of the ether extract between petroleum ether and aqueous methanol gave a slight but still insufficient improvement.

An attempt was made to develop a quicker and cheaper method for sample purification than that used for solid tissue but with a noise value further below the expected levels of contamination. The analytical strategy was based on steric exclusion chromatography, a method much used in isolation and purification procedures for steroids (Ager &

Oliver, 1983). A separation medium based on cross-linked styrene-divinyl benzene was selected, since this can be used at high solvent pressures and gives a more efficient and quicker separation than the more commonly employed cross-linked agarose. HPLC-based systems are also easier to automate. As will be seen, this choice was fortuitous. The full procedure was as follows. Hydrolysis with glucuronidase/sulphatase and ether extraction were carried out as previously mentioned. A partition between petroleum ether and aqueous methanol was included for reasons which will become apparent later. The most important step of the method was, however, that involving the steric exclusion chromatography (SEC). This uses 30 cm columns packed with beads of cross-linked styrene-divinyl benzene and elution with toluene at a flow rate of 1 ml/min.

The particular packing used was Ultrastyragel (Waters Associates) with 100 Å pore size. The unexpected bonus found with this phase is shown in Fig. 4 which demonstrates the nature of the separation of the anabolic hormones. Trenbolone is clearly separated from the others and zeranol is adequately separated from the stilbenes. Since all these compounds are of closely related molecular weight, steric exclusion may not

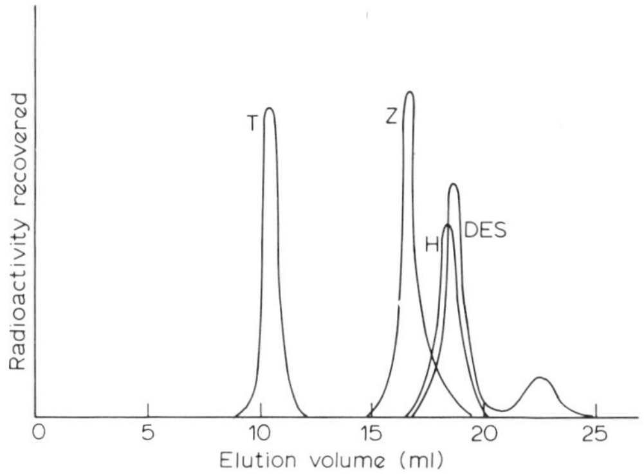

Fig. 4. Elution of tritiated hormones from column of 100Å cross-linked styrene–divinyl benzene (Ultrastyragel, Waters Associates). T, Trenbolone; Z, zeranol; H, hexoestrol; DES, diethylstilboestrol.

be the only mechanism operating. The use of toluene containing 5% tatrahydrofuran (THF) results in all the hormones being eluted together in about 8 ml. The considerable adsoption of zeranol and the stilbenes may possibly be due to their aromaticity or acidity. There is no significant cross-reactivity of any of the anabolic steroids with antisera of the other groups; therefore they could all be eluted together with toluene/THF and then separated for their different immunoassays. However, the separation using toluene/THF results in poorer clean up of the sample; it would also involve using larger volumes of bile, giving rise to practical difficulties associated with larger volumes of solvent and reduced throughput.

It will be seen from Fig. 4 that ^{3}H-DES gives two elution peaks on this particular column. It is not known whether the smaller second peak represents an impurity in the ^{3}H-DES or, possibly, separation of *cis* and *trans* isomers since it is present in all tritiated preparations we have received; in any case, no cross reactivity with antisera 6139 and 254 is associated with the material in the second smaller peak. As the separation medium in the column ages, the separation of the smaller peak from the main DES peak becomes progressively poorer. The first peak, which contains the bulk of the DES and hexoestrol, is routinely taken for analysis.

Figure 5 again shows the effect of bile on the standard curve for stilbenes using antiserum 6139, but this time after subjecting the bile extract

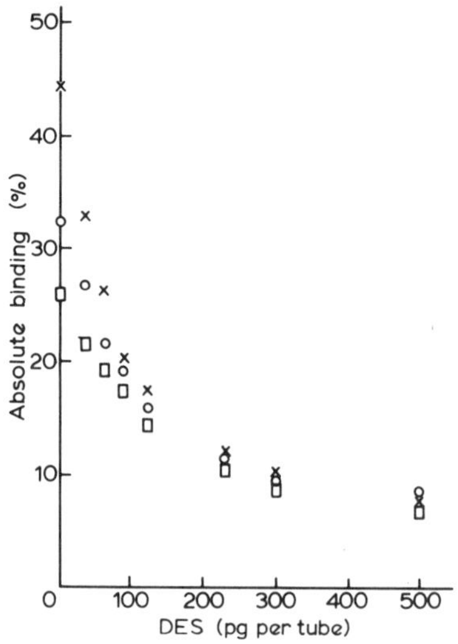

Fig. 5. Standard curves using DES and antiserum 6139. (i) ×, Normal standard curve; (ii) □, standard curve in presence of 50 μl ether extracted bile subjected to steric exclusion chromatography on Ultrastyragel; (iii) ○, as in (ii) but with aqueous methanol/petroleum ether partition before being applied to column. All results corrected for NSB.

to SEC chromatography. There is a distinct improvement in the shape of the standard curve over that obtained without chromatography. It can also be seen that the inclusion of the extra partition step itself gives a significant further improvement compared to the standard curves containing extracted bile shown in Fig. 3. The same pattern of results was found for antiserum 254 (DES specific).

Figure 6 shows the results obtained versus the results expected for DES standards in the presence of fully cleaned-up bile extracts. The line indicates the ideal result. The results indicate a positive bias from the ideal for both antisera but this is reasonably consistent throughout the

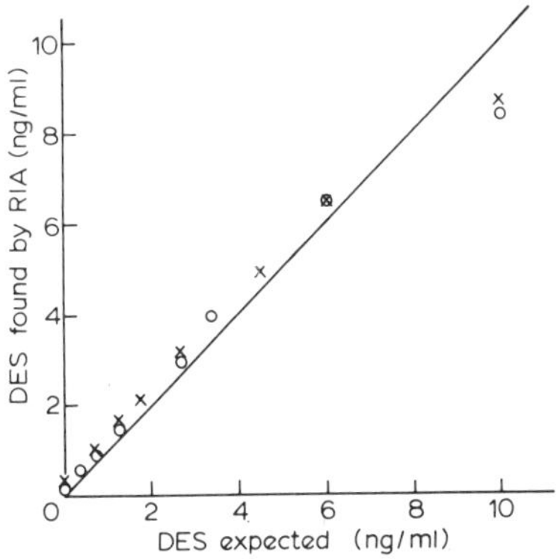

Fig. 6. Results from bile extracts spiked with DES standards, versus expected result. ×, Using antiserum 6139; ○, using antiserum 254; line represents ideal result.

range of the curve up to 6 ng/ml. The deviation from the line at the 10 ng point is consistent with it being on the limit of the standard curve where, typically for this non-linear dose response assay, the coefficient of variation is highest (Malvano, 1982). This procedure has been used routinely for the analysis of stilbenes in bile for over a year, so that reliable basic data are now available. These are shown for antiserum 6139 in Table 4. The results clearly indicate that the tissue background value (process noise) is well below the expected analytical signal (cf. Table 1). These results have been corrected for solvent blank, then converted to pg/ml; apparent negative results are due to subtraction of the solvent blank. Since using this procedure, our cost per sample has been reduced to less than £20 and throughput has increased to approximately 40 samples per person per week.

Table 4
Basic Data on the Analysis for Stilbenes in Veal Bile using Antiserum 6139

	Mean	*SD*	*Range*	*n*
Blank (pg)	9	8	1–37	18
Tissue background (pg/ml)	−14	189	−360 to 854	221
Spike (pg/ml)	10 583[a]	3 336	4 343–15 818	18

[a]Should be 16 000

The effect of bile extracts on the zeranol radio-immunoassay is less marked than for stilbenes, though the results are still inadequate. A significantly lower background value, and a usable assay, is obtained with the full sample purification procedure, when compared with a simple ether extract of bile. α-Trenbolone analysis is the least satisfactory with this purification technique, the full clean-up procedure as described still results in a considerable non-parallelism and high tissue background value. Subsequent treatment of the column eluent on small alumina columns gives a distinct improvement, but there is still more scatter than obtained with the other hormones and higher solvent blank values. However, it should be possible, with further development, to make this assay for α-trenbolone in bile suitable for routine use. If this proves to be the case, then it will be possible to monitor the use of all the major synthetic anabolic hormones on one extract (if triplicate analysis was carried out, less than 400 μl of bile would suffice).

AN ALTERNATIVE APPROACH

The results so far presented have not given a full picture of the effects of bile on the assays and the benefits of the methodological improvements, since they have all had non-specific binding values (NSBs) deducted. Values in Table 5 show that NSBs are usually greatly increased in the presence of bile and bile extracts, presumably by interfering in the charcoal separation stage, and that the purification procedure eliminates this interference. From this table, it can again be seen that the other effect of bile is depression of the absolute binding figure, which is only partly relieved by the purification procedure. Much of the remaining depression of binding is due to solvent effects and, on subtraction of this, the residual gives rise to the reported tissue background values. With the

Table 5
Typical Variations of Absolute Binding (AB) and Non-specific Binding (NSB) with Bile Clean-up Procedure (as Percent of Total Radioactivity Added)

	Stilbenes				*Zeranol*		*α-Trenbolone*	
	6139		*254*					
	AB(%)	*NSB*(%)	*AB*(%)	*NSB*(%)	*AB*(%)	*NSB*(%)	*AB*(%)	*NSB*(%)
1. Normal standard curve	55	10	61	12	69	2	70	4
2. Direct bile addition (50 μl)	—	—	—	—	—	—	12	75
3. After ether extraction	11	25	15	45	31	5	21	24
4. After full clean-up	45	14	51	12	54	2	38	5

information in Table 5 it became possible for us to adopt another approach to our assay problems.

The approach outlined above takes existing immunoassay methodologies and develops purification procedures to allow extracts of tissues to be used in them. A more satisfactory, though longer-term approach, is to develop immunoassays to suit the system to be measured and the circumstances in which it is to be used. The strategy for this is as follows. Firstly, ours is a specialised central laboratory with the equipment and experience to use radiotracers. A statistically valid sampling scheme demands that samples are sent for analysis from all over the country, which is both costly and inefficient. In order to allow analyses to be carried out regionally, some other more convenient end-point detection method is required; this can conveniently be achieved by making use of ELISA (enzyme-linked immunosorbent assay) techniques. Secondly, even ELISA would be of little value if purification procedures were complex or used complex equipment. Therefore an assay is required in which bile can be used directly.

Table 5 clearly shows the two major problems; firstly, the elevated NSBs which are presumably due to interference with the separation stage (charcoal). A better approach would be to use a solid-phase separation technique, e.g. one reacting component being adsorbed to the walls of the reaction vessel. Separation of the other component which has not bound to the first one at the end of the reaction is then achieved simply by aspiration and washing. Secondly, type II assays (competitive, using a limited amount of antibody) are especially susceptible to compounds which, acting at other than the binding site, affect the kinetics and equilibrium of the assay. In contrast, type I (non-competitive) assays, in which excess antibody is used, and analyte concentration is determined after separation of free and bound antibody, should be less susceptible to such compounds since excess antibody 'drives' the analyte to complex formation (Ekins, 1981). It is possible that bile constituents, in addition, may act at other than the analyte binding site, so the type I non-competitive approach should prove more useful in this particular situation.

The following assay procedure is based on that used by Dr M. R. A. Morgan of the Food Research Institute, Norwich. The monovalent antigen is first coupled to a carrier protein and adsorbed onto the walls of the wells of a microtitre plate (Fig. 7). Excess antibody is equilibrated with the sample and the reaction mixture then transferred to the plate. The free antibody, not complexed with antigen in the sample, will

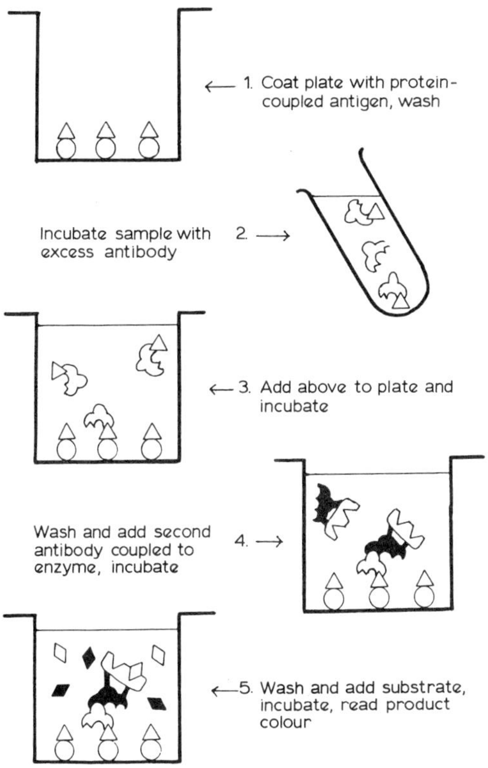

Fig. 7. Outline of a non-competitive ELISA method.

combine with the adsorbed, conjugated antigen and remains in the well when the reaction medium is aspirated off at the end of the first incubation period on the plate. An enzyme-labelled second antibody, raised against IgG from the species used to raise the primary antiserum, is added and binds in direct proportion to the number of primary antibodies present. After a further period of incubation, the excess labelled second antibody is removed by aspiration and after washing the plate, enzyme substrate is added. The enzyme on the second antibody catalyses a colorimetric reaction, the absorbance being directly proportional to the amount of primary antibody unused in the initial equilibration, i.e. inversely proportional to the concentration of analyte in the sample.

A typical standard curve achieved by this method, using 17α-trenbolone, is seen in Fig. 8; also shown in the same curve in the presence of 50 μl bile (*not* extracted). Clearly, using this approach, the NSB is not affected by the presence of bile, there is, though, still some depression of

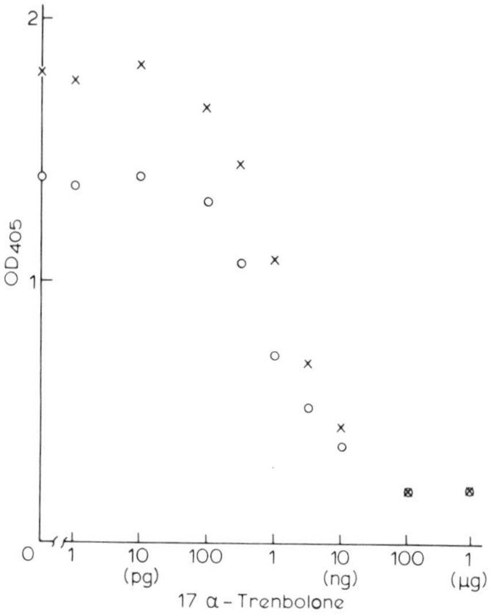

Fig. 8. Standard curve for analysis of 17 α-trenbolone using a non-competitive ELISA. ×, Normal standard curve; ○, standard curve in presence of 50 μl bile.

binding. However, these results should be compared with those in Fig. 9, obtained by radioimmunoassay, using the same antiserum and coated charcoal to effect separation of free and bound analyte. The radioimmunoassay exhibits very high non-specific binding and greatly depressed absolute binding in the presence of 50 μl bile (*not* extracted). The use of a type I immunoassay (ELISA) as opposed to type II immunoassay (RIA) clearly brings about significant improvement and there seems no reason why manipulation of the system should not allow it to be improved further.

This chapter has highlighted some of the problems encountered in applying immunoassays to trace analysis in complex food systems. The first, more obvious approach, of attempting to purify samples can often produce a usable method in a short time, but may ultimately be self defeating as more and more purification steps are required. We would

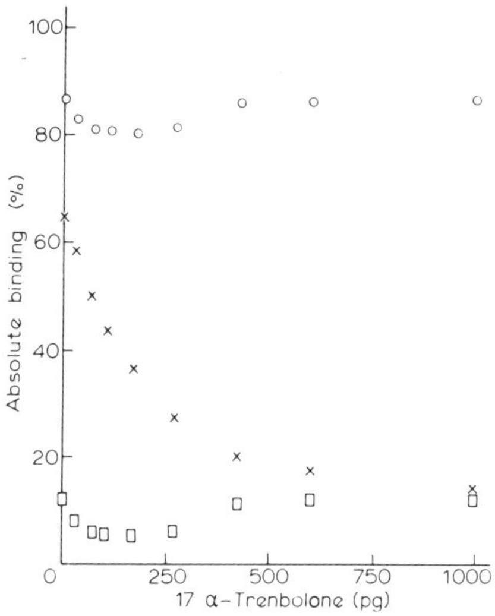

Fig. 9. Standard curve for analysis of 17 α-trenbolone using a RIA method. (i) ×, Normal standard curve; (ii) ○, standard curve in presence of 50 μl bile; (iii) □, as in (ii) with NSB subtracted.

recommend that the second approach, using a non-competitive assay system, should also be considered since it makes use of the great versatility of immunoassay in tailoring an assay to individual requirements.

REFERENCES

AGER, R. P. & OLIVER, R. W. A. (1983) Separation of oestrogens in biological fluids and synthetic mixtures on Sephadex G-type gels. *Journal of Chromatography*, **271**, 325–40.

EEC (1981*a*) EC Directive (81/602/EEC) on the marketing and use of anabolic growth promoters and thyrostatic agents. O. J. L222, 7.8.81, HMSO, London.

EEC (1981*b*) Radioimmunoassay methods for the measurement of residues of the stilbene derivatives, diethylstilboestrol, hexoestrol and dienoestrol in tissues and fluids of farm animals. Directorate — General for Agriculture, 4694/VI/81-EN, File No. 67.1, Commission of the European Communities, Brussels.

EKINS, R. (1981) Merits and disadvantages of different labels and methods of immunoassay In: *Immunoassays for the 80s*, Voller, A., Bartlett, A. & Bidwell, D. (eds), MTP Press Ltd, Lancaster, UK, pp. 5–16.

HARWOOD, D. J., HEITZMAN, R. J. & JOUQUEY, A. (1980) A radioimmunoassay method for the measurement of residues of the anabolic agent, hexoestrol in tissues of cattle and sheep. *Journal of Veterinary Pharmacology and Therapeutics*, **3**, 245–54.

HEITZMAN, R. J. & HARWOOD, D. J. (1983) Radioimmunoassay of hexoestrol residues in faeces, tissues and body fluids of bulls and steers. *Veterinary Record*, **112**, 120–3.

HOFFMANN, B. (1978) Use of radioimmunoassay for monitoring hormonal residues in edible animal products. *Journal of the Association of Official Analytical Chemists*, **61** (5), 1263–73.

IARC (1979) *Monographs on the Evaluation of the Carcinogenic Risks of Chemicals to Humans*, Vol. 21, *Sex Hormones* (11), International Agency on Cancer, Lyon, France.

LU, F. C., RENDEL, J., COULSTON, F. & KARTE, F. (eds) (1976) Proceedings of the FAO/WHO Symposium *Anabolic Agents in Animal Production*. Rome, 1975. In: *Environmental Quality and Safety*, Supplement Vol. V, Georg Thieme Publishers, Stuttgart.

MALVANO, R. (1982) Error in RIA: Sources and evaluation. *Journal of Nuclear Medicine and Allied Sciences*, **26** (4), 205–26.

MARKS, V. (1981) Immunoassays in pharmacology and toxicology. In: *Immunoassays for the 80s*, Voller, A., Bartlett, A. & Bidwell, D. (eds), MTP Press Ltd, Lancaster, UK, pp. 263–75.

14

Comparison of the Analysis of Total Potato Glycoalkaloids by Immunoassays and Conventional Procedures

M. R. A. MORGAN, R. MCNERNEY, D. T. COXON and H. W.-S. CHAN

AFRC Food Research Institute, Norwich, UK

PROPERTIES OF POTATO GLYCOALKALOIDS

Potato glycoalkaloids are a class of nitrogen-containing steroidal glycosides commonly found in plants of the Solanum genus, including the cultivated potato, *Solanum tuberosum*. They consist of either a spirosolane- or solanidane-type aglycone, to which up to four carbohydrate molecules are attached. In potatoes, the two major glycoalkaloids which make up 95–99% of the total glycoalkaloid (TGA) content of tubers (Pasechnichenko & Guseva, 1956) are α-solanine and α-chaconine, both solanidine triglycosides (Fig. 1). TGA levels vary within the potato, and, in the tuber, highest concentrations are found in the peel immediately beneath the skin and in areas of high metabolic activity, such as eye regions (Wolf & Duggar, 1940; Lampitt *et al*., 1943). Cooking and other types of processing do not appear to reduce potato TGA levels (Baker *et al*., 1955; Bushway & Ponnampalam, 1981).

Several reports have linked high potato TGA levels with cases of poisoning in both animals and man (for review, see McMillan & Thompson, 1979). Symptoms include severe gastro-intestinal upset, cardiovascular and respiratory depression, leading occasionally to death. The glycoalkaloids inhibit cholinesterase activity (Orgell *et al*., 1958) and have been implicated as teratogens in several animal species (Mun *et al*., 1975; Keeler *et al*., 1976; Allen *et al*., 1977). Elevated levels in the potato have been shown to impart an unpleasant bitter flavour to the cooked product (Sinden & Deahl, 1976). For these reasons, commercially available potatoes should contain below 20 mg TGA/100 g, a level generally regarded as being safe for consumption.

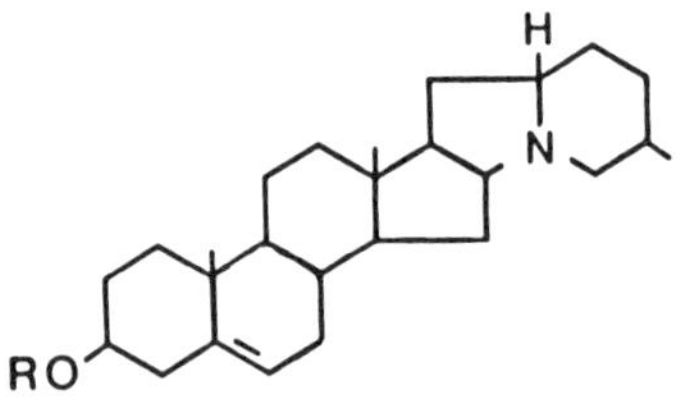

R = H — Solanidine

R = Gal-Glu
    Rham — Solanine

R = Glu-Rham
    Rham — Chaconine

Fig. 1. Structure of the glycoalkaloids α-solanine and α-chaconine, and of the aglycone solanidine (gal = galactose; glu = glucose; rham = rhamnose).

TGA levels in tubers depend on three factors:

(i) cultivation factors, such as soil type, geographical location of growth, tuber age and fertilization practice;
(ii) post-harvest treatment and storage, including mechanical or chemical induced stress of the tuber and exposure to light;
(iii) the genetic background of the tuber.

Under normal circumstances, potatoes currently available in the UK have been shown to contain between 1 and 15 mg TGA/100 g. There does not, therefore, seem to be any cause for concern as long as continued monitoring of TGA levels is used to check changing growing and marketing practices for existing varieties. However, the situation is slightly different as regards potential new potato varieties. The potato breeder is looking for greater pest and disease resistance in new lines. Unfortunately, it appears that these desirable qualities are sometimes linked with unacceptably high tuber TGA levels. All putative new varieties need to be checked at an early stage in the development and testing programme to avoid costly later withdrawal. At least one such action has occurred — the cultivar Lenape had to be withdrawn shortly after its introduction in the USA (Zitnak & Johnston, 1970).

There is clearly a requirement for suitable routine methods of analysis for TGA determination.

CONVENTIONAL METHODS OF TGA ANALYSIS

Conventional methods of total glycoalkaloid analysis have recently been extensively reviewed (Coxon, 1984). Two of the most widely used and accepted methods are those of Sanford & Sinden (1972) and Fitzpatrick & Osman (1974). The method of Sanford and Sinden involves a lengthy extraction procedure and a colorimetric reaction with antimony trichloride and concentrated hydrochloric acid prior to TGA quantification. Disadvantages of this reliable method are the time taken for each analysis and the very unpleasant nature and toxicity of the colorimetric reagents. It will not measure Δ 5-unsaturated glycoalkaloids, such as demissine.

The Fitzpatrick and Osman method employs a bisolvent extraction of TGA, followed by phase separation. The TGA-containing solution is then taken to dryness and hydrolysed with sulphuric acid to yield the aglycones. After re-extraction, the aglycones are titrated against bromophenol blue and quantified. This much faster method, requiring no specialised equipment, has been criticised for lack of accuracy and reliability (Butcher, 1978; Coxon *et al.*, 1979; Mackenzie & Gregory, 1979). The main problem appears to be the low and variable recoveries encountered even with standards in the absence of potato. 'Recovery factors' have been employed to correct for these problems.

As well as many published modifications to these methods, other types of assay have been described, though less widely used. Thin-layer chromatographic methods have the advantages of simplicity and potential cheapness, but are difficult to use quantitatively. Gas-chromatographic procedures for glycoalkaloids require analyte derivatisation, and are often carried out under conditions that give very short column lives — a distinct disadvantage where routine analysis is required. High performance liquid chromatography has the potential for fast and accurate quantification; however, the requirement for expensive equipment and the lack of a suitable glycoalkaloid chromophore have limited wide application of this technique.

IMMUNOASSAY METHODS FOR TGA ANALYSIS

Immunoassay methods are sensitive, specific and well-suited to routine application, particularly where large sample numbers are involved. These properties are the consequence of the robust affinity and specificity of an antibody for its antigen. Two immunoassays have been described for the determination of TGA in potato. The method of Vallejo & Ercegovich (1979) was a radioimmunoassay for solanidine. Antisera were raised against solanidine by injecting rabbits with a solanidine–bovine serum albumin conjugate, synthesised by making the hemisuccinate derivative of the aglycone and coupling it to the protein by the mixed anhydride method. As a consequence, the antiserum cross-reacted comparatively poorly (at only 28% of the solanidine binding) with α-chaconine and α-solanine. It was therefore necessary to convert the glycoalkaloids to aglycones by acid hydrolysis prior to radioimmunoassay of potato extracts. Correlation of radioimmunoassay results for TGA analysis in tubers with those obtained by the Fitzpatrick & Osman (1974) method was good. Disadvantages of the immunoassay were the requirement for the 2 h hydrolysis stage, the need for radio-labelled solanidine, which is unavailable commercially, and for access to expensive liquid scintillation counting facilities, itself a time-consuming procedure. In addition, the anti-solanidine antiserum used in the assay was used at a final working dilution of only 1:25 (v/v). Such a poor titre immediately precludes wider use of the assay in other laboratories.

At the Food Research Institute, Norwich, we have preferred to develop enzyme immunoassays. For the food analyst, these offer a cheaper and more flexible alternative, and without the problems associated with the handling and disposal of radioactivity. Many of the recent developments in enzyme immunoassay ensure that their performance can be at least equal to that of radioimmunoassays. The particular type of enzyme immunoassay on which we have concentrated has been the micro-titre plate enzyme-linked immunosorbent assay (ELISA), which offers particular advantages of simplicity and potential for automation

Our approach to setting up an immunoassay for potato glycoalkaloids has been to generate a broad specificity antiserum so as to avoid having to use a hydrolysis step. Thus α-solanine has been coupled to bovine serum albumin, through the sugar residues, by a periodate cleavage method. This conjugate has been used to raise anti-glycoalkaloid anti-sera in rabbits, giving a product that recognises α-chaconine, α-solanine and demissine — the predominant potato glycoalkaloids — equally well. In the ELISA for potato TGA determination (Morgan *et al.*, 1983), the potato material is simply homogenised and diluted prior to assay. No hydrolysis is required and recovery is quantitative. The antiserum is used at dilutions of up to 1:20 000 (v/v).

ANALYSIS OF POTATO TGA BY ELISA AND A CONVENTIONAL METHOD

(i) Potato Samples

Potatoes were obtained from the Potato Department of the Scottish Crop Research Institute, Pentlandfield, Midlothian. These were products of the breeding programme and are not commercially available. A representative sample of five washed tubers (100–150 g each) was selected for each variety. Two opposite eighth sections from each tuber were cut out and chopped into small pieces. This material was either analysed immediately or frozen in liquid nitrogen and stored at −20°C.

(ii) ELISA Analysis

The ELISA analysis of potato samples was carried out as previously described (Morgan *et al.*, 1983).

The assay is based on the varying amounts of rabbit anti-glycoalkaloid antibody that will bind to immobilised glycoalkaloid in the presence of varying amounts of glycoalkaloid in solution. Thus, the less glycoalkaloid in solution (in the sample or standard), the more anti-glycoalkaloid antibody will bind to the immobilised phase. After washing away free material, these differing amounts of primary antibody are measured by adding an anti-rabbit IgG antibody to which an enzyme has been covalently bound. After washing again, the amount of primary antibody present is determined by adding enzyme substrate. Quantification of glycoalkaloid in the original sample is by measurement of optical density and reference to standard curves.

Potato (5 g) was prepared by homogenisation (2 min) in 50 ml of a methanol–water–acetic acid (94:6:1) mixture. After allowing insoluble debris to settle (5 min), an aliquot of the supernatant was removed and diluted as required in assay buffer. No correction was made for recovery, since this has always been measured at greater than 90%.

Assays were carried out in the wells of microtitre plates, previously coated with glycoalkaloid–protein conjugate to form the immobilised phase. These coated plates are stable for at least nine months when stored dry at room temperature. Anti-glycoalkaloid antiserum (available from the AFRC, Food Research Institute, Norwich, UK) was employed at a 1:3000 dilution (v/v), and anti-rabbit IgG antiserum, conjugated to alkaline phosphatase (Miles Labs. Ltd, Slough, UK), was diluted 1:1000 (v/v). After substrate incubation, well optical densities at 405 nm were recorded.

(iii) Analysis by Sanford and Sinden Method

Duplicate samples of potato (20 g) were extracted for 16 h in a Soxhlet apparatus with a solvent mixture consisting of ethanol (150 ml) and glacial acetic acid (3 ml), following the procedure of Sanford & Sinden (1972). After precipitation, washing and drying, the purified glycoalkaloid precipitates were dissolved in ethanol:2N HCl (95:5) and determined colorimetrically, using the antimony trichloride reagent described by Wierzchowski & Wierzchowska (1961), modified by the addition of ascorbic acid (Bretzloff, 1971). A calibration curve was constructed using purified α-chaconine as a standard.

(iv) Results

The sensitivity of the ELISA method under the conditions employed was 0·15 mg TGA/100 g potato tissue, and that of the chemical method 0·25 mg/100 g. Inter- and intra-assay coefficients of variation were below

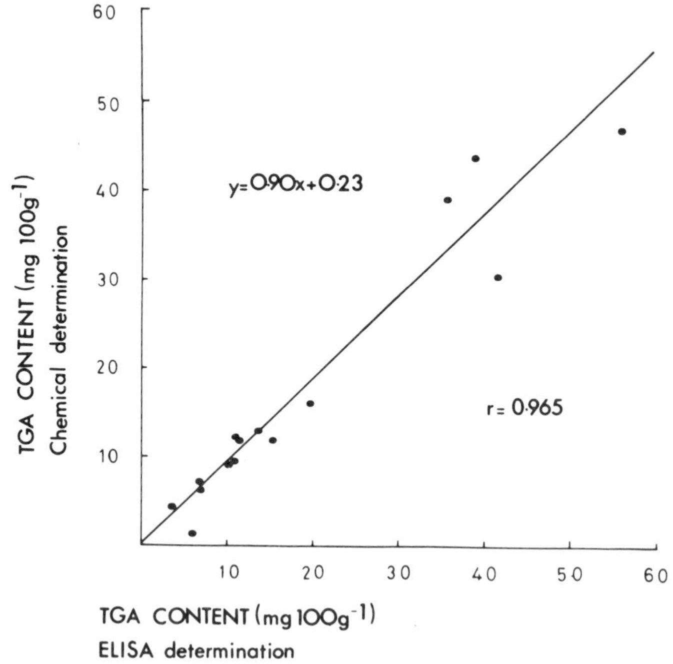

Fig. 2. Correlation of results for potato TGA obtained by the chemical method and by the ELISA technique ($n = 15$). See text for full details.

10% for both methods for tuber samples of low, medium and high TGA content in the range studied.

Figure 2 shows the results obtained for each potato sample by the chemical method plotted against the result obtained by ELISA. The correlation coefficient, *r*, is 0·965 for the 15 samples.

(v) Comparison of Methods

Sensitivity is not usually a problem in the analysis of potato TGA. Indeed, the ELISA is operating at well below its potential, being able to assay as little as 2 μg TGA/100 g. The sensitivity of the immunoassay technique now enables solanidine to be measured routinely in human plasma for the first time (Matthew *et al.*, 1983), an advance that should yield useful information about the metabolism of this potentially toxic compound.

The specificity of the ELISA and Sanford and Sinden methods is such that current varieties can be assayed perfectly well. There may be a problem for both methods with possible new varieties of unusual genetic background. The Sanford and Sinden method does not assay for demissine and similar Δ 5-unsaturated glycoalkaloids. The ELISA does measure these solanidane-based compounds, but will not assay spirosolane-derived materials, such as tomatine. It must be said that this problem would be anticipated to be an extremely rare one. These assays cover more than 95% of the normal TGA fraction. For a problem to occur, not only would new glycoalkaloids have to be present in considerable quantity, but also the existing glycoalkaloids would have to remain at acceptably low levels.

The correlation of results obtained by the two methods, based on different principles, seems to be excellent. The disadvantages of the chemical method are, however, clear. The reagents can be extremely unpleasant and toxic. The method requires a high degree of analytical skill and is very tedious. The samples in the present study were assayed at the rate of nine per week, a rate not sufficient to cope with routine monitoring. In contrast, the ELISA method employs stable, non-toxic reagents. Technical skill is limited to the ability to pipette accurately and repetitively. No specialised equipment is required apart from a means of measuring optical densities, and inexpensive portable equipment is now commercially available for this purpose. The number of potato samples that can be assayed routinely is in excess of 250 per week.

In conclusion, conventional methods of potato TGA analysis have problems and disadvantages, particularly for routine monitoring purposes. The ELISA method described overcomes these drawbacks and produces results in close agreement with alternative procedures. The ELISA would appear to be the method of choice for potato TGA analysis.

ACKNOWLEDGEMENTS

The skilful assistance of Sara Turner is gratefully acknowledged.

REFERENCES

ALLEN, R. J., MARLAR, R. J., CHESNEY, G. F., HELGESON, J. P., KELMAN, A., WECKEL, K. G., TRAISMAN, E. & WHITE, J. W. (1977) Teratogenicity studies on late blighted potatoes in non-human primates (*Macaca mulatta* and *Saquinus labiatus*). *Teratology*, **15**, 17–23.

BAKER, L. C., LAMPITT, L. H. & MEREDITH, O. B. (1955) Solanine, glycoside of the potato. III. An improved method of extraction and determination. *Journal of the Science of Food and Agriculture*, **6**, 197–202.

BRETZLOFF, C. W. (1971) A method for rapid estimation of glycoalkaloids in potato tubers. *American Potato Journal*, **48**, 158–62.

BUSHWAY, R. J. & PONNAMPALAM, R. (1981) α-Chaconine and α-solanine content of potato products and their stability during several modes of cooking. *Journal of Agricultural and Food Chemistry*, **29**, 814–7.

BUTCHER, H. (1978) Total glycoalkaloids and chlorophyll in potato cultivars bred in New Zealand. *New Zealand Journal of Experimental Agriculture*, **6**, 127–30.

COXON, D. T., (1984) Methodology for glycoalkaloid analysis. *American Potato Journal*, **61**, 169–83.

COXON, D. T., PRICE, K. R. & JONES, P. G. (1979) A simplified method for the determination of total glycoalkaloids in potato tubers. *Journal of the Science of Food and Agriculture*, **30**, 1043–9.

FITZPATRICK, T. J. & OSMAN, S. F. (1974) A comprehensive method for the determination of total potato glycoalkaloids. *American Potato Journal*, **51**, 318–23.

KEELER, R. F., BROWN, D., DOUGLAS, D. R., STALLKNECHT, G. F. & YOUNG, S. (1976) Teratogenicity of the *Solanum* alkaloid solasodine and of Kennebec potato sprouts in hamsters. *Bulletin of Environmental Contamination and Toxicology*, **15**, 5221–5.

LAMPITT, L. H., BUSHILL, J. H., ROOKE, H. S. & JACKSON, E. M. (1943) Solanine, glycoside of the potato. 2. Distribution in the potato plant. *Journal of the Society of Chemistry and Industry, London*, **62**, 48–51.

MACKENZIE, J. D. & GREGORY, P. (1979) Evaluation of a comprehensive method for total glycoalkaloid determination. *American Potato Journal*, **56**, 27–33.

McMillan, M. & Thompson, J. C. (1979) An outbreak of suspected solanine poisoning in schoolboys: examination of criteria of solanine poisoning. *Quarterly Journal of Medicine*, **48**, 227–43.

Matthew, J. A., Morgan, M. R. A., McNerney, R., Chan, H. W.-S. & Coxon, D. T. (1983) Determination of solanidine in human plasma by radioimmunoassay. *Food and Chemical Toxicology*, **21**, 637–40.

Morgan, M. R. A., McNerney, R., Matthew, J. A., Coxon, D. T., & Chan, H. W.-S. (1983) An enzyme-linked immunosorbent assay for total glycoalkaloids in potato tubers. *Journal of the Science of Food and Agriculture*, **34**, 593–8.

Mun, A. M., Barden, E. S., Wilson, J. M. & Hogan, J. M. (1975) Teratogenic effects in early chick embryos of solanine and glycoalkaloids from potatoes injected with late blight *Phytophthora infestans*. *Teratology*, **11**, 73–8.

Orgell, A. H., Vaidya, K. A. & Dahm, P. A. (1958) Inhibition of human plasma cholinesterase *in vitro* by extracts of solanaceous plants. *Science*, **128**, 1136.

Pasechnichenko, V. & Guseva, A. R. (1956) Quantitative determination of potato glycoalkaloids and their preparative separation. *Biochemistry (USSR)*, **21**, 606–11.

Sanford, L. L. & Sinden, S. L. (1972) Inheritance of potato glycoalkaloids. *American Potato Journal*, **49**, 209–17.

Sinden, S. L. & Deahl, K. L. (1976) Effect of glycoalkaloids and phenolics on potato flavour. *Journal of Food Science*, **41**, 520–3.

Vallejo, R. P. & Ercegovich, C. D. (1979) Analysis of potato for glycoalkaloid content by radioimmunoassay (RIA). *Trace Organic Analysis: A New Frontier in Analytical Chemistry, Proceedings of the 9th Materials Research Symposium*, National Bureau of Standards (USA), Special Publication 519, pp. 333–40.

Wierzchowski, P. & Wierzchowska, Z. (1961) Colorimetric determination of solanine and solanidine with antimony trichloride. *Chemio Analityczna Warszawa*, **6**, 579–85.

Wolf, M. J. & Duggar, B. M. (1940) Solanine in the potato and the effects of some factors on its synthesis and distribution. *American Journal of Botany*, **27** Supplement 20s.

Zitnak, A. & Johnston, G. R. (1970) Glycoalkaloid content of B5141-6 potatoes. *American Potato Journal*, **47**, 256–60.

15

Cross-reactions in Immunoassays for Small Molecules: Use of Specific and Non-specific Antisera

R. J. Robins, M. R. A. Morgan, M. J. C. Rhodes and J. M. Furze

AFRC Food Research Institute, Norwich, UK

INTRODUCTION

Immunoassays provide an analytical system with which trace amounts of a desired compound may be assayed with little or no requirement for purification and concentration of the sample. This is made possible by the highly specific nature of the antibody–antigen interaction and the extreme sensitivity with which interaction may be observed. In many applications of the technique this high specificity is a desirable property. In this presentation, a case is made that, in certain situations, it is beneficial deliberately to develop an immunoassay with decreased specificity. Furthermore, such a development need not lead to loss of sensitivity.

At the AFRC Food Research Institute in Norwich, immunoassays are being developed for the analysis of a number of compounds of low molecular mass (< 1000 daltons). The use of these assays falls into two major categories, the first of which contains a number of assays for the detection of toxic compounds at trace levels in foods. The second category is in association with work on plant cell cultures, where an experimental programme is currently being followed to boost the level of the production of useful secondary metabolites in certain species. An immunoassay for these metabolites makes it possible to detect them in very small samples of tissue. In both categories, examples arise in which the extreme sensitivity of an immunoassay is desirable, but where the target is to detect a group of metabolites, rather than individual molecular species. The examples cited show how, by judicious manipulation of the nature of the antigen, it is possible to induce a population of antibodies which possess these properties.

In the conventional immunogenic response against a purified protein, a population of antibodies is elicited specific to its surface features (i.e. the antigenic determinants). The nature of the antigenic determinants is thus predetermined by the structure of the protein being investigated. When raising antibodies to a low molecular weight compound the immunogen is a protein to which the small molecule has been covalently bound. The antibody population generated should thus include antibodies possessing a recognition site capable of interacting with the small molecule, even when this is not bound to the protein. This has been found in the several hundred cases so far tested. When the small molecule is linked to the protein, close proximity of the hapten to the protein is required to elicit antibody production. Such a structure means that a region of the small molecule close to the protein is masked by the polypeptide chain, thus playing no role in the determination of the conformation of the recognition site on the antibody. Hence, by linking a small molecule through different moieties on its structure, specific or non-specific antibodies may be generated, depending on the distribution within the compound of other variable functional groups relative to the point of conjugation.

FOOD TOXINS

The first example for consideration in this category is the antiserum raised to ochratoxin A (Fig. 1) and used to develop an ELISA for this compound (Morgan *et al.*, 1983*a*). This potent mycotoxin, produced by a number of species of the fungal genera *Aspergillus* and *Penicillium*, can become a serious contaminant in a range of foodstuffs, notably cereals, and may be the cause of fatal human kidney diseases (Hartwig, 1974). Ochratoxin A is usually found associated with traces of its ethyl ester, ochratoxin C (although this may be an artifact of the extraction process), and rather more ochratoxin **B** (5-dechloro-ochratoxin A; Fig. 1), which is not implicated as a toxin. Hence, an assay is required which will determine the toxic ochratoxins A and C in the presence of relatively large amounts of the non-toxic ochratoxin B.

Of the several possible functional groups available for conjugation, the carboxylic acid residue was chosen as this is distal to the C-5 position at which the distinctive chlorine atom occurs and leaves free both ring

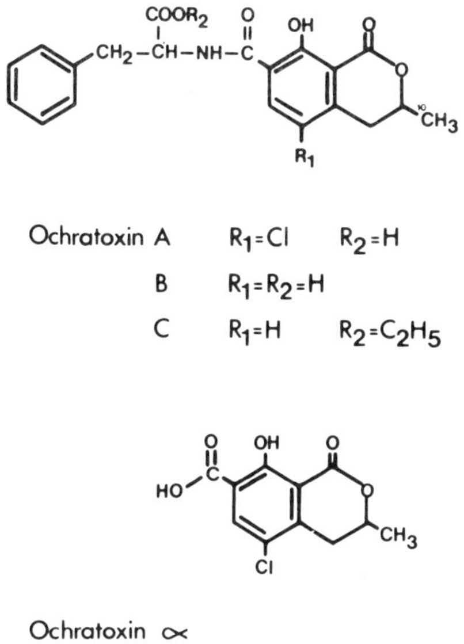

Fig. 1. Chemical structures of ochratoxins.

structures on the molecule. Linkage was readily achieved by the mixed anhydride method (Erlanger *et al.*, 1959). The free carboxylic acid residue is reacted (Fig. 2) with iso-butylchlorocarbonate under alkaline conditions to form an acid anhydride intermediate, which readily couples to the free amine groups of the protein, in this case bovine serum albumin, to form a suitable immunogen.

When used in an ELISA, the antiserum was shown to have the predicted properties. It is of high sensitivity, detecting ochratoxin A over the range 5–1000 pg, and of extreme specificity, showing a cross-reaction with ochratoxin B of only 0·5% (Table 1). Hence only the toxin will be detected in foodstuffs. Animals can, to some extent, detoxify ochratoxin A either by hydroxylation (Stormer *et al.*, 1983) — as in so many detoxification mechanisms — or by cleavage of the peptide bond, yielding phenylalanine and ochratoxin α (Storen *et al.*, 1982). Hydroxylation in the 4R position leads to a product giving only a 7% cross-reaction, while the other known metabolites which cross-react at all only do so at 1–2% (Table 1). The assay is thus also valuable for detecting ochratoxin A in tissues from animals. A particular example of

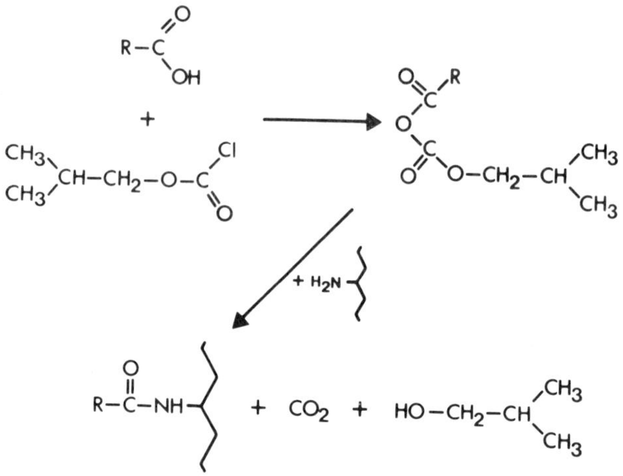

Fig. 2. Scheme showing the coupling of a compound containing a carboxylic acid residue (R-COOH) to a protein using the mixed anhydride method of Erlanger *et al.* (1959).

Table 1
Cross-reactivity of Anti-ochratoxin A Antiserum

Compound	*Cross-reaction* (%)
Ochratoxin A	100
Ochratoxin B	0·5
Ochratoxin C	100
Ochratoxin α	2·4
(4R)-4-Hydroxyochratoxin A	1·3
(4S)-4-Hydroxyochratoxin A	7·2
10-Hydroxyochratoxin A	1·4
Phenylalanine	0

this application is the need to examine kidneys from pigs suspected of having been fed on contaminated feed (Hartwig, 1974).

In direct contrast to this assay is the example of the ELISA developed for the determination of total glycoalkaloids in potato tubers (Morgan *et al.*, 1983*b*). These steroidal alkaloids can be toxic at very low concentrations and cases have been reported linking their consumption with poisoning and death in humans and animals (McMillan & Thompson, 1979). Their association with bitterness helps, to some extent, to alleviate the problem (Sinden & Deahl, 1976).

In typical, commercially available potato varieties, 95% of the total glycoalkaloid occurs as glycosylated solanidine, either as α-solanine or as α-chaconine (Pasechnichenko & Guseva, 1956), differing only in the composition of the tri-saccharide substituent (Fig. 3). All three com-

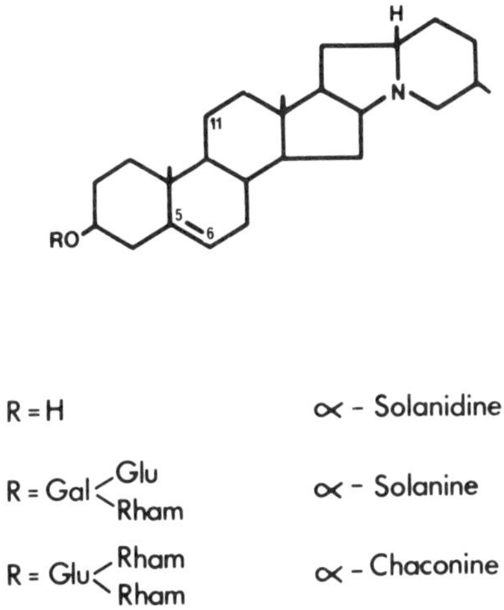

Fig. 3. Chemical structures of glycoalkaloids.

pounds shown in Fig. 3 are toxic (Nishie *et al.*, 1971) and therefore an immunoassay was required which would detect them equally well. It was argued that, by linking to protein through the sugar residues then a conjugate should be formed capable of eliciting antibodies unable to distinguish α-solanine from α-chaconine or from the aglycone, as the variable residues would be destroyed by the conjugation reaction.

An immunogen was therefore synthesised by periodate treatment of α-solanine (Fig. 4) which leads to cleavage of the C—C bonds between adjacent hydroxyl groups on the sugars, leaving aldehyde residues. These readily react with the free amino groups of bovine serum albumin under alkaline conditions to form a Schiff base, which on reduction with sodium borohydride leads to a stable immunogen (Butler & Chen, 1967).

With this antiserum, an identical dose–response curve is obtained

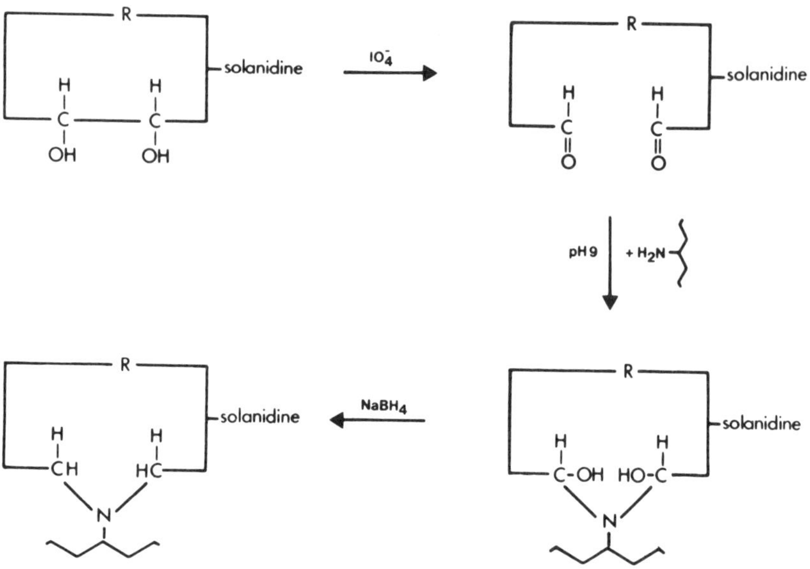

Fig. 4. Scheme showing the coupling of solanidine through its sugar residues (R—CH—CH with OH OH) using the periodate cleavage method of Butler & Chen (1967).

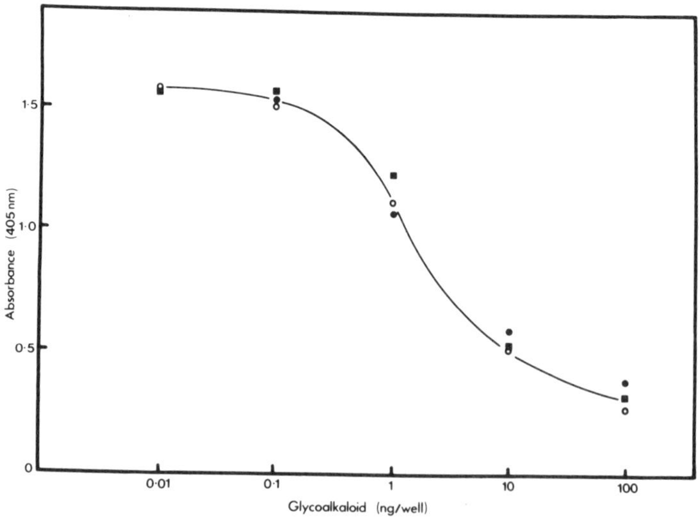

Fig. 5. Standard curves for α-solanidine (○). α-solanine (●) and α-chaconine (■), in the ELISA for the determination of total alkaloids in potato tubers as described by Morgan *et al.* (1983*b*).

whether α-solanine, α-chaconine or α-solanidine is assayed (Fig. 5). The detection range is 0·1–10 ng. When cross-reactions are examined with other molecules (Table 2), the effect of changes close to the site of conjugation is exemplified by demissidine, which is 5,6-dihydrosolanidine (Fig. 3) and not distinguished from solanidine despite the effect that this reduction will have on the overall stereochemistry of the free molecule. A small alteration in the distal part of the molecule, however, has a major effect, as is seen with rubijervine, which is 11-hydroxysolanidine (Fig. 3), and only cross-reacts at 4%. Several other alkaloids, including solasodine, show no cross-reaction (Table 2). It is important to note that neither do the phytosterols, β-sitosterol and stigmasterol, both likely to be present in crude extracts of potato at far higher concentrations than the glycoalkaloids.

Table 2
Cross-reactivity of Anti-α-solanine Antiserum

Compound	*Cross-reaction* (%)
α-Solanine	100
α-Chaconine	100
α-Solanidine	100
Demissidine	100
Rubijervine	4·0
Solasodine	0
β-Sitosterol	0
Stigmasterol	0

PLANT METABOLITES

The second application of immunoassays to be considered involves their use for the detection of secondary metabolites in plant cell cultures. A major restriction to progress towards realizing the potential that plant cell cultures have for the production of secondary products is that their synthesis tends to occur at only very low levels in cultured cells. The difficulty involved in stimulating the cells to elevate their production suggests that the relevant genetic material is severely repressed. While one approach towards alleviating this problem is to use biochemical manipulation, an attractive alternative is to screen the population and select out the very small number of cells present which are naturally derepressed. This latter approach requires the analysis of samples from many thousands of microcultures derived by cell plating, a task only feasible using immunoassay.

At present, we are studying the synthesis of quassin (Fig. 6), the major bitter seco-triterpene produced by the tropical tree *Quassia amara* (Simaroubaceae). In addition to quassin, *Q.amara* synthesises a group of precursors and metabolites, notably neoquassin, 18-hydroxyquassin and

Class 1

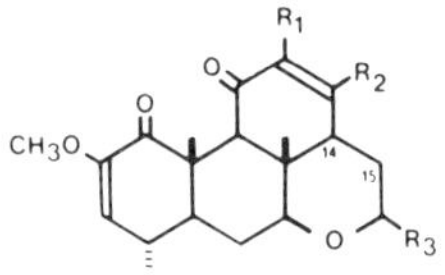

		R_1	R_2	R_3
I	Quassin	OCH_3	CH_3	=O
II	Neoquassin	OCH_3	CH_3	HOH
III	12-hydroxyquassin	OH	CH_3	=O
IV	18-hydroxyquassin	OCH_3	CH_2OH	=O
V	14,15-dehydroquassin	OCH_3	CH_3	=O

Class 2

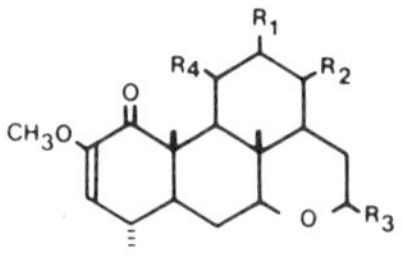

		R_1	R_2	R_3	R_4
VI	Nigakilactone A	OH	CH_3,H	=O	OH
VII	Nigakilactone B	OCH_3	CH_3,H	=O	OH
VIII	Nigakilactone E	OCH_3	CH_3,OH	=O	OAc
IX	Nigakilactone F	OCH_3	CH_3,OH	=O	OH
X	Nigakihemiacetal A	OCH_3	CH_3,OH	H,OH	OH

Class 8

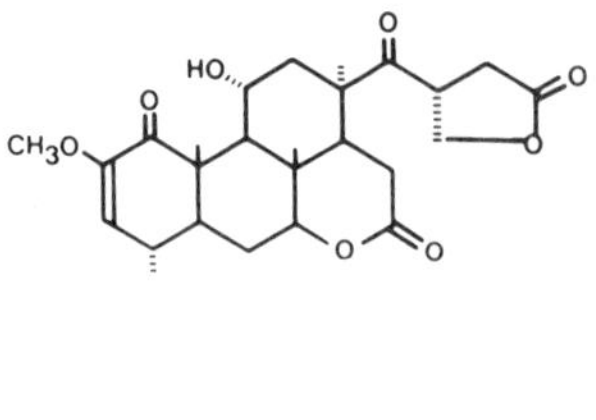

XX Picrasin A

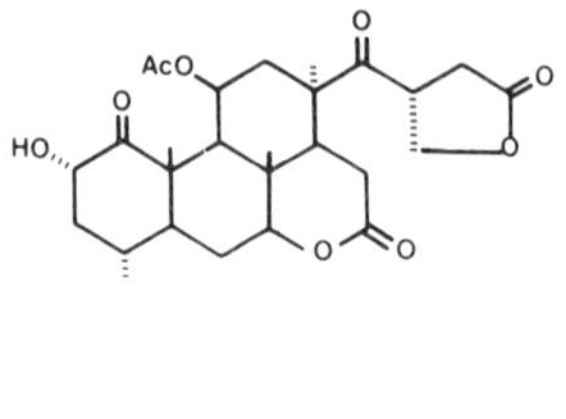

XXI Simarolide

Fig. 6. Chemical structures of quassinoids.

12-hydroxyquassin (Fig. 6), all of which are of interest. In all these molecules, the A and B rings are structurally identical, modifications being confined to the C and D rings. An antiserum was required which was able to detect all four of these products.

Conjugation was therefore achieved, in the first instance (Robins *et al.*, 1984*a*) by opening the lactone ring, isomerising the molecule to stabilise the carboxylic acid product and coupling to bovine serum albumin by the mixed anhydride method (Erlanger *et al.*, 1959; Fig. 2). Commercially available quassin actually contains about 25% neoquassin, 4% 18-hydroxyquassin and 1% 12-hydroxyquassin (Robins *et al.*, 1984*a*). Deliberately, this preparation was not purified further so that the two hydroxyl-substituted forms would also be conjugated with protein. Thus, an immunogenic mixture of a defined nature was generated.

The antiserum obtained was tested for cross-reaction with 25 naturally occurring quassinoids (Table 3 and Fig. 6) and found to have the desired

Table 3
Cross-reactivity of the Anti-isoquassinic Acid Antiserum

Compound	*Cross-reaction* (%)
Quassin	100
α-Neoquassin	36
β-Neoquassin	100
12-Hydroxyquassin	100
18-Hydroxyquassin	250
14,15-Dehydroquassin	81
Iso-quassinic acid methyl ester	64 000
Nigakilactone A	33
Nigakilactone B	11
Nigakilactone F	6·9
Nigakihemiacetal A	15
Picrasin A	13·5
15 other quassinoids	$0{\cdot}5 > x \geqslant 0$

features (Robins *et al.*, 1984*a*). As predicted, it shows great susceptibility to any changes to the distal face of the molecule, in particular, in the A ring, but is relatively unresponsive to changes at the proximal faces of the C and D rings. But, in this instance, producing an antiserum of low selectivity has also led to decreased sensitivity and the limit of detection is only 5 ng. This is not due to a poor quality antiserum. The working dilution used is 1:80 000 and standard curves can be obtained at even greater dilutions. Furthermore, if the methyl ester of isoquassinic acid (in which the structure of the antigen is maintained) is examined, then a detection range of 5–100 pg is observed, comparable to many other systems. Relative to this compound, all cross-reactions are at less than 0·1%.

For some purposes, a higher sensitivity antiserum was desirable. To synthesise this, 18-hydroxyquassin was isolated by preparative high performance liquid chromatography (Robins & Rhodes, 1984) and used for immunogen synthesis (Robins *et al.*, 1984*b*). The 18-hemisuccinate derivative was prepared (Fig. 7) by reacting the 18-hydroxyl group with succinic anhydride in pyridine. This was then linked to bovine serum albumin by the mixed anhydride technique involving isobutylchlorocarbonate (Erlanger *et al.*, 1959). The stereochemistry of 18-hydroxyquassin is extremely similar to quassin, as shown by ^{1}H-NMR studies (R. J.

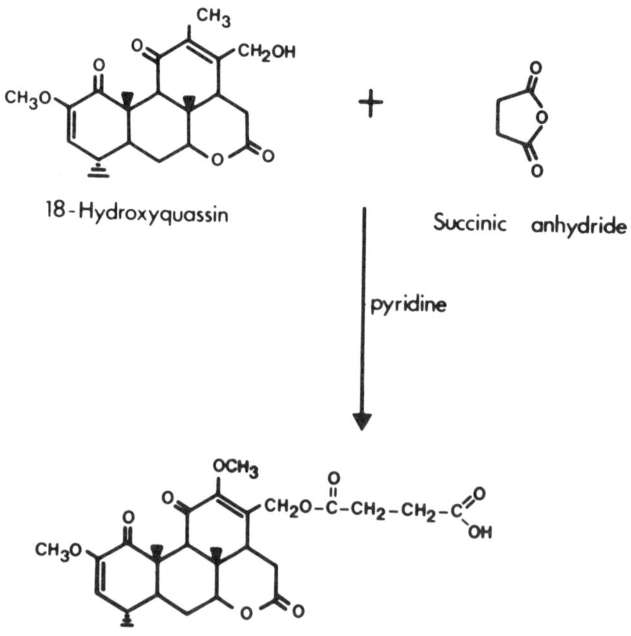

Fig. 7. Scheme showing the synthesis of 18-hydroxyquassin hemisuccinate as described by Robins *et al.* (1984*b*).

Robins & D. T. Coxon, unpublished results), and should mimic quassin precisely as an antigen when conjugated in this manner.

As predicted, when tested for cross-reactions, the antiserum was unable to distinguish between quassin and 18-hydroxyquassin (Table 4).

Table 4
Cross-reactivity of the Anti-18-hydroxyquassin Antiserum

Compound	*Cross-reaction* (%)
18-Hydroxyquassin	100
Quassin	89
α-Neoquassin	11·5
β-Neoquassin	189
12-Hydroxyquassin	128
14,15-Dehydroquassin	10·7
Nigakilactone A	34
Nigakilactone B	45
Nigakilactone F	2·3
Nigakihemiacetal A	1·0
14 other quassinoids	$0{\cdot}5 > x \geqslant 0$

Furthermore, with a dose–response curve over the range 5–500 pg, it proved to be of comparable sensitivity to ELISAs for ochratoxin A and potato glycoalkaloid. No distinction was made between 12-hydroxyquassin and quassin, but steric changes in the molecule, as in 14,15-dehydroquassin (Fig. 6) or isoquassinic acid, caused considerable loss of recognition. Indeed, these molecules were recognised less well than some of the nigakilactones (Fig. 6), quassinoids synthesised by another member of the family, *Picrasma quassioides*. Changes to the proximal side have little effect on cross-reaction, but alterations on the distal side have a major effect on recognition.

The final example to be considered is not of an ELISA but a RIA (Robins *et al.*, 1984*c*) for another bitter compound, quinine (Fig. 8), an alkaloid produced by tropical trees of the genus *Cinchona* (Rubiaceae). In this case, the cross-reactivity measured in the RIA was precisely as predicted from a study of the stereochemistry of the structures concerned.

Four major quinoline alkaloids (Fig. 8) are made by *Cinchona* species, another six occurring in only trace quantities. These ten compounds fall

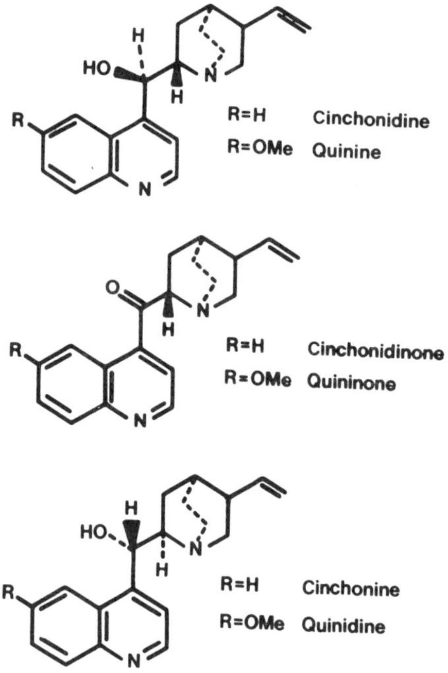

Fig. 8. Chemical structures of *Cinchona* alkaloids.

into two series which, although biosynthetically closely related, have differing stereochemistries and fundamentally different three-dimensional configurations. In cell cultures, all four products are made, but only one of these, quinine, is of interest to the food industry. Another, quinidine, is of interest to the pharmaceutical industry. It is therefore desirable to be able to screen cultures separately for quinine and its precursor cinchonidine, which have 8S,9R stereochemistry, and for quinidine and its precursor cinchonine, of 8R,9S configuration (Fig. 8). The only free hydroxyl group in these compounds is at C-9 and conjugation at this point, with only a four-carbon bridge, should cause the antigens to be presented to solution on the surface of the protein in a way that will show their maximal steric differences due to the C-8 stereochemistry.

Conjugation of quinine was achieved, as with 18-hydroxyquassin, by reaction with succinic anhydride to form a hemisuccinate (Fig. 9) and coupling this to bovine serum albumin with isobutylchlorocarbonate (Erlanger *et al.*, 1959). A radiolabel for RIA was prepared by acetylating

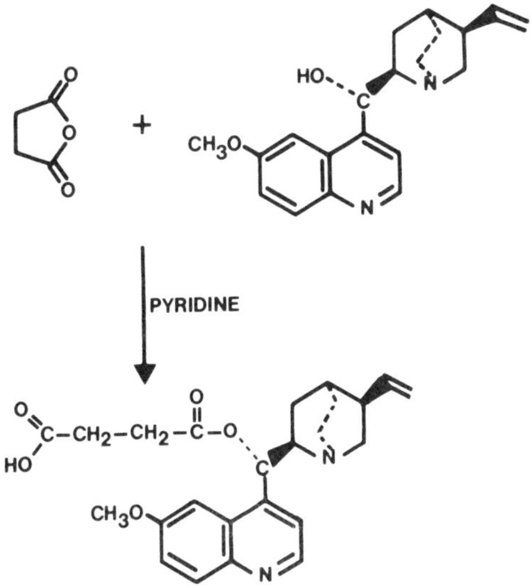

Fig. 9. Scheme showing the synthesis of quinine-9-hemisuccinate for conjugation to bovine serum albumin using the mixed anhydride method of Erlanger *et al.* (1959).

the hydroxyl with ^{3}H-acetic anhydride. The anti-quinine antiserum shows exactly the results predicted (Table 5). No cross-reaction at all is found with the 8R isomers quinidine and cinchonine (Fig. 8). Neither are quininone or cinchonidinone, molecules in which the C-9 is oxidised to a ketone causing the relative orientation of the two rings to be in-

Table 5
Cross-reactivity of the Anti-Quinine Antiserum

Compound	*Cross-reaction* (%)
Quinine	100
9-Acetylquinine	329
10,11-Dihydroquinine	35
Cinchonidine	14
Quininone	0
Cinchonidinone	0
Quinidine	0
10,11-Dihydroquinidine	0
Cinchonine	0

termediate to the 8S,9R and 8R,9S series, recognised. Significant cross-reaction only occurs with 10, 11-dihydroquinine and with cinchonidine, a precursor of quinine lacking the methoxyl substituent at C-6′. Thus, the assay will detect quinine and its precursor cinchonidine over a range of 50 pg–100 ng in samples which may also contain the stereoisomers of these products. The antiserum is thus ideally suited to perform the task for which it was designed. An antiserum of similar properties has been developed for the measurement of quinidine using an ELISA (Morgan *et al.*, 1985).

CONCLUSION

Non-specific and specific antisera are both shown to be useful in food analysis. In particular, a role is demonstrated for the use of non-specific antisera to determine the presence of groups of compounds, a technique of considerable potential. Within the examples cited, two ways are described in which non-specific antisera may be obtained. In one, the variant parts of the group of compounds of interest are masked by conjugation through the residue in which variation occurs; in the other a defined mixture of immunogens is used to raise an antibody population capable of tolerating small changes within the antigen, but sensitive to alterations within specified regions of the molecule. By prior careful consideration of the molecular structures involved, it is shown possible to raise antibodies able only to recognise one stereoisomer of a pair or, in contrast, a general shape containing a number of surface features of key interest.

ACKNOWLEDGEMENTS

Our thanks are due to Ruth McNerney, Jennifer Matthew and Andrew Webb who carried out much of the technical aspects of the work reported here.

REFERENCES

Butler, V. P. & Chen, J. P. (1967) Digoxin-specific antibodies. *Proceedings of the National Academy of Sciences*, USA, **57**, 71–8.

ERLANGER, B. F., BOREK, O. F., BEISER, S. M. & LEIBERMAN, S. (1959) Steroid–protein conjugates II. Preparation and characterization of conjugates of bovine serum albumin with progesterone, deoxycorticosterone and estrone. *Journal of Biological Chemistry*, **234**, 1090–4.

HARTWIG, J. (1974) Ochratoxin A and related metabolites. In: *Mycotoxins* Purchase, I. F. H. (ed), Elsevier, Amsterdam, pp. 345–67.

MCMILLAN, M. & THOMPSON, J. C. (1979) An outbreak of suspected solanine poisoning in schoolboys. *Quarterly Journal of Medicine*, **48**, 227–43.

MORGAN, M. R. A., MCNERNEY, R. &. CHAN, H. W.-S. (1983*a*) The enzyme-linked immunosorbent assay of ochratoxin A in barley. *Journal of the Association of Official Analytical Chemists*, **66**, 1481–4.

MORGAN, M. R. A., MCNERNEY, R., MATTHEW, J. A., COXON, D. T. & CHAN, H. W.-S. (1983*b*) An enzyme-linked immunosorbent assay for total glycoalkaloids in potato tubers. *Journal of the Science of Food and Agriculture*, **34**, 593–8.

MORGAN, M. R. A., TURNER, S., WEBB, A. J., ROBINS, R. J. & RHODES, M. J. C. (1985) Specific immunoassays for quinine and quinidine: comparison of radioimmunoassay and enzyme-linked immunosorbent assay procedures. *Planta medica* (in press).

NISHIE, K., GUMBMANN, M. R. & KEYL, A. C. (1971) Pharmacology of solanine. *Toxicology and Applied Pharmacology*, **19**, 81–92.

PASECHNICHENKO, V. & GUSEVA, A. R. (1956) Quantitative determination of potato glycoalkaloids and their preparative separation. *Biochemistry (USSR)*, **21**, 606–11.

ROBINS, R. J. & RHODES, M. J. C. (1984) High performance liquid chromatographic methods for the analysis and purification of quassinoids from *Quassia amara* L. *Journal of Chromatography*, **283**, 436–40.

ROBINS, R. J., MORGAN, M. R. A., RHODES, M. J. C. & FURZE, J. M. (1984*a*) An enzyme-linked immunosorbent assay for quassin and closely-related metabolites. *Analytical Biochemistry*, **136**, 145–56.

ROBINS, R. J., MORGAN, M. R. A., RHODES, M. J. C. & FURZE, J. M. (1984*b*) Determination of quassin in picogram quantities by an enzyme-linked immunosorbent assay. *Phytochemistry*, **23**, 1119–23.

ROBINS, R. J., WEBB, A. J., RHODES, M. J. C., PAYNE, J. & MORGAN, M. R. A. (1984*c*) Radioimmunoassay for the quantitative determination of quinine in cultured plant tissues. *Planta medica*, **50**, 235–8.

SINDEN, S. L. & DEAHL, K. L. (1976) Effects of glycoalkaloids and phenolics on potato flavour. *Journal of Food Science*, **41**, 520–3.

STOREN, O., HOLM, H. & STORMER, F. C. (1982) Metabolism of ochratoxin A by rats. *Applied and Environmental Microbiology*, **44**, 785–9.

STORMER, F. C. STOREN, O., HANSEN, C. E., PEDERSEN, J. I. & AASEN, A. J. (1983) Formation of (4R)- and (4S)-4-hydroxyochratoxin A and 10-hydroxyochratoxin A from ochratoxin A by rat liver microsomes. *Applied and Environmental Microbiology*, **45**, 1183–7.

Index

ABEI, 78
Accelerators, of precipitate formation, 43–4
Adulteration of meat, detection of, 87–94, 95–109
Affinity, xv, 38, 118, 173; *see also* Avidity
Affinity chromatography, 61, 89, 97–100, 104, 127, 147–9; *see also* Immunoadsorbent chromatography
Aflatoxin, 20
Agar gel immunodiffusion test (AGDT), 95–6, 103, 106–7
AHEI, 78
Albumin, serum, 38, 88–93
  bovine, 38, 89–93, 110, 116, 122, 127, 136–7, 163, 171, 190, 199, 201, 205–6, 208–9
  horse, 89–92
  sheep, 89–93
Aldehyde residue, 201
Alkaline phosphatase, 56, 59–60, 62, 103, 116, 127, 136, 143, 145–6, 191
Amatoxins, 20
Amino groups; *see* Lysine, ε-amino groups
Ammonium sulphate accelerator, 44
Ammonium sulphate precipitation, 40, 41, 96, 99, 132
Amylase, 125, 129, 138
Amyloglucosidase, 121–38
Anabolic hormones, 169–85
  separation by steric exclusion chromatography, 177
  structures, 170
Anabolic steroids; *see* Anabolic hormones
Analysis, bio-ochratoxins, 160–1
Analysis, chemical
  glycoalkaloids, 189, 192–3
  ochratoxins, 160–1
Analysis, immuno-, 111
  barley, ochratoxins, 162–5, 198–200
  beer, amyloglucosidase, 126–38
  bile, anabolic hormones, 171–80, 182–5
  cell cultures, quassins, 205–7
    quinine alkaloids, 207–10
  meat, anabolic hormones, 171–80, 182–5
    species identification, 88–93, 96–102, 102–7
  milk proteins, 116–21
  porcine kidney, ochratoxin, 162–5, 198–200
  potato, glycoalkaloids, 190–4, 200–3
  protein denaturation, 116–21
  soya protein, 111–13
  staphylococcal enterotoxins, 142–52
Analyte, xv, 6, 22, 54, 57, 64, 76, 88, 172–3, 175, 182, 184

Anaphylaxis in guinea pigs, 115–16, 121–2
Androgens, 169; *see also* Anabolic hormones
Antibiotics, 20, 67
Antibodies
  affinity purified, 89, 98, 104
  ammonium sulphate isolation, 96
  analytes, as, 58, 67, 78, 95–108
  enzyme-labelled, affinity purified, 138
  immunoglobulins, to, 90
  monoclonal, xix, 49, 90, 104, 107–8, 151–2, 172
  monospecific, 89–90, 93, 103–4, 107
  non-precipitating, 21
  precipitating, 21, 87, 103
    disadvantage in ELISA, 103
  rivanol isolation, 96
  species-specific, 88–90, 97–8, 102, 104
Antibody, xv
  binding sites, 22, 34, 171, 182, 198
  Fab fragment, 96–101, 144, 146
  Fc fragment, 42–3, 58, 144, 146
  labelled, 22, 46–9, 75
  primary, also antibody, first, xvii, 162–3, 183, 191
  production, 28–37, 96–7, 116, 198
    species of animal for, 30–1, 145
  second, xx, 31, 41–4, 45–6, 49, 90, 144, 161
    labelled, 49; *see also* Enzyme-labelled second antibody
    solid-phased, 45–6
  specificity, phylogenetic relationships, 89–93
  water of hydration, 44
Antigen, xv
  labelled, xix, 21, 22, 24–6, 53–67, 73–82
  meat species, 96
  monovalent, 182
  multivalent, xix, 57
Antigen-free matrix, xv, 38, 165, 172, 176, 178–9
Antigenic binding site, 48
Antigenic determinant, xv, 48–9, 57, 89, 100, 123, 144–5, 151, 198
Antigenicity,
  guinea pigs, in, 115–16, 121–3
  protein, *in vitro* test, of, 115–23
  reduction of, by heat treatment, 115–16
Antiserum, xv
  assessment, 31–7
    avidity, 33–4
    specificity, 34–7
    titre, 31–3
  dilution curve, 31–3, 64
  displacement curve, 32–3
  monospecific; *see* Antibodies, monospecific
  non-specific, 197–210
  polyclonal, xx, 88–9, 107, 127, 131, 138
  primary; *see* Antibody, primary
  production; *see* Antibody production
Aryl acridinium esters, as labels, 81–2
Aspergillus, 159, 198
Aspergillus niger, 126, 129
Aspergillus oryzae, 127, 129, 131
Assay conditions, 37–8, 131–5
Association time, of antibody, 27–8
Avidin, as label in EIA, 152
Avidity, xv, 26–8, 31, 33–4, 41, 43, 48, 64–5, 107

Baby milk formulas, 115
Bacteriophage, 20
Bacteriostat, 38, 59, 62
Barley, ochratoxin analysis, 164
Beef, detection of, 87, 89–93, 95, 100–2, 104–7
Beer
  amyloglucosidase analysis, 126–39
  analysis of, 20
  low-calorie, production of, 126
Bias, xvi, 179
Bile
  stilbene analysis, 176–80
  trenbolone analysis, 182–5
Bilirubin, effect of, 74
Bioluminescence, 78–9
Bioluminescent labels, 25

Biotin, as label in EIA, 152
Bitter principles, analysis for, 20
Boar taint, analysis for, 20
Bound fraction, xvi
Bridge recognition, 26, 65–6, 206
Buffalo meat, detection of, 95, 101–2, 104–7

Caffeine, analysis for, 20
Calibration curve; *see* Standard curve
Camel meat, influence of, 105
Capture, immunoassay, 102–7
Carbodiimide, 61, 78, 161–3
Carboxylic acid residue, 198–200, 205–6
Carrier proteins, xvi, 161–3, 171, 182, 198–9, 201, 205–6, 208–9; *see also* Albumin, serum, bovine; Egg albumen; Immunoglobulin G, bovine; Keyhole limpet haemocyanin; Lysozyme; Poly-L-lysine
Carrier serum, 43
Casein, α_{sl}, 116, 121, 123; *see also* Milk proteins
Cattle meat, detection of, 100–2; *see also* Beef, detection of; Veal, detection of
Cellulose, microcrystalline, 45, 63
Cereals, analysis of, 20
  ochratoxin A analysis, 164, 198
α-Chaconine, 187–8, 190, 201–3; *see also* Glycoalkaloids
Chemiluminescence, 78–82, 152
Chemiluminescence immunoassay, 81, 142
Chemiluminescent labels, 25, 77–82
Chloramine-T, 25
Chromatoelectrophoresis, 39–40
Chromatography,
  immunoadsorbent, 38, 47, 88–90, 93, 104
  steric exclusion, 176–9
Ciguatoxin, analysis for, 20
Cinchona alkaloids, 207–10; *see also* Quinine alkaloids
Cinchonidine, 208–10; *see also* Quinine alkaloids
Cinchonidinone, 208–9; *see also* Quinine alkaloids
Cinchonine, 208–9; *see also* Quinine alkaloids
Classical immunoassay, xvi, 21–46
Clostridium toxins, analysis for, 20
Coated charcoal, 41–2, 45, 171–2, 180, 182, 184
Coated tube, 40, 44–5, 62–3, 66, 142
Coenzymes, as labels, 74
Collaborative trial, 113
Colostral whey Ig, 97
Competitive ELISA, xvi, 56–8, 142–4, 162
Complement, 43
β-Conglycinin, 120
Conjugation methods; *see* Carbodiimide; Diazotisation; Glutaraldehyde; Hetero-bifunctional reagents; *N*-Hydroxysuccinimide ester; Mixed anhydride reaction; Periodate oxidation
Coprecipitation, 40
Cross-reaction, xvi, 33–5, 90, 104, 107, 129, 138, 171, 177–8, 197–210
  with metabolites, 163, 172
Cross-reactivity,
  abrogation of, 104
  antisera to, of,
    amyloglucosidase, 129, 138
    anabolic hormones, 171–2
    glycoalkaloids, 190, 193, 203
    meat species, 90–2, 103–7
    milk protein, 120–1
    ochratoxins, 164, 199–200
    quassins, 205–7
    quinine alkaloids, 209
    small molecules—rationale, 197–210
    soya products, 112

Deer meat, influence of, 90, 92
Dehydroquassin, 205, 207; *see also* Quassins

Delayed addition immunoassay, xvi
Demissine, 188–190, 193, 203; *see also* Glycoalkaloids
Detection limit, xvii, 27, 31, 64, 141, 148, 161–5, 175, 192–3, 205
Detergent, 20, 66, 90, 98, 105, 111, 128
effect of, 134–5, 137
Dextran-coated charcoal, 41, 171–2, 180, 182
Diazotisation, 78
Dienostrol, 171–2
Diethylstilboestrol (DES), 170–2, 174, 176–9; *see also* Anabolic hormones
Differentation of
heat processed and cooked meats, 93
raw, unheated meats, 87–93
Dihydroquinine, 210; *see also* Quinine alkaloids
Disequilibrium immunoassay, xvii
Donkey meat, detection of, 100–1
Donkeys, for antiserum production, 31
Double antibody, xx, 33, 41–4, 62–3, 162; *see also* Antibody, second
Double immunodiffusion, detection limits, 141
Drugs, assays for, 76

EDTA, 43
Egg albumen, 161
ELISA, competitive inhibition method for milk proteins, 115–18, 123
ELISA, correlation with other analytical methods, 105–7, 113, 121–3, 148, 150, 164, 184–5, 192–3
ELISA, enzyme inhibitors in, 163
ELISA, indirect, 57–8, 60, 145–6, 162–3
ELISA, non-specific interference in, 112, 163
ELISA, protein A interference in, 146
Enterotoxin A, 141–3, 145–6
Enterotoxin B, 142, 145–51
Enterotoxin C, 142, 145–6
Enterotoxin D, 141, 145
Enterotoxin E, 145
Enterotoxins, 141–52
Enthalpy, 65
Enzymes as analytes, 20
amyloglucosidase, 121–39
esterase, 104, 106
isoenzyme staining, 104, 106
lactate dehydrogenase, 104, 106
Enzymes as labels, 25, 26, 58–60, 103; *see also* Alkaline phosphatase; β-D-Galactosidase; Horseradish peroxidase; Urease
Enzyme immunoassay, xvii, 53–67, 96, 102–8, 142, 190
capillary tubes, in, 108
capture, 102
magnetic inhibition, 147–8
Enzyme-labelled antibody, 56–7, 127–8
Enzyme-labelled antigens, 40, 42
Enzyme-labelled anti-species IgG antibody, *see* Enzyme-labelled second antibody
Enzyme-labelled conjugates, 60–1
Enzyme-labelled haptens, 55, 61–2, 65–6, 162
Enzyme-labelled proteins, 61
Enzyme-labelled reagents, 53, 55, 58–62, 74
Enzyme-labelled second antibody, 57–8, 60, 90, 102, 116–17, 145–6, 183
advantages of, 162–3
Enzyme-linked immunosorbent assay (ELISA), xvii, 21, 22, 36, 49, 55–60, 63–7
amyloglycosidase in beer, 126–38
glycoalkaloids in potato, 190–4, 200–3
milk protein denaturation, 116–21
ochratoxin A in food, 162–5, 198–200
soya protein, 111–13
species identification of raw, unheated meat products, 88–93, 102–7
staphylococcal enterotoxins in food, 142–52
trenbolone in bile, 182–5

Enzyme mediated immunoassay technique (EMIT), xvii, 55, 57, 147
Enzyme promotors, in ELISA, 163
Equilibrium immunoassay, xvii
Europium chelate, as label, 77
Excess reagent immunoassay, xvii, 46–9, 57, 64, 182

Fab fragment, 96–101, 107, 144, 146
Fc fragment, 42–3, 58, 144, 146
Fat, analysis of, 20
First antibody; *see* Antibody primary
Fish, analysis of, 20
Fluoroimmunoassay, xvii, 74–7, 152
  enhancement, 76
  polarisation, 76
  quenching, 75–6
  release, 76
  time resolved, 77
Fluorophores, 74–6
  endogenous, 74
  inner filter effects, 74
  light scattering effects, 74–5
  quenching, 74
Fluoroscein, as label, 74–6
Food analyte, distinction from traditional analyte, 6
Food immunoassay, 3–19
  bibliography, 8–19
  definition, 3
Food poisoning, 141
Food products, analysis of
  canned, 111
  dried, 111
  heat set, 111
  raw, 111
Food toxins, 198–203
Free fraction, xvii, 22
Free radicals, as labels, 74
Freund's complete adjuvant, 97
Fruit and fruit products, analysis of, 20
Functional groups, 198; *see also* Aldehyde residue; Amino groups; Carboxylic acid residue; Hemisuccinate; Hydroxyl groups

β-D-Galactosidase, 56, 59–60, 62, 76
Gelatin, 38, 98
γ-Globulins, 42; *see also* Immunoglobulin
Glutaraldehyde, 61, 127, 147
Glycoalkaloids, 20, 187–94, 200–3
  chemical analysis, 189
  immunoanalysis, 191–3, 202, 203
  structure, 187–8, 201
Goat anti-rabbit IgG, 90, 116–17, 145
Goat meat, detection of, 90–3, 95, 103–6
Goats, for antiserum production, 88, 90, 116–17, 145–6
Ground beef, 105
Guinea pigs, antigenicity in, 115–16, 121–2

Haemoglobin, 74
Hapten, xviii, 26, 31, 53, 55, 61–2, 65, 66, 75–6, 162, 198–210
Hapten–enzyme conjugates; *see* Enzyme-labelled haptens
Heat stable proteins, 93
Heat treatment of proteins, to reduce antigenicity, 115–16, 119, 121, 123
Hemisuccinate, preparation, 206, 208–9
Herbicides, assays for, 67
Hetero-bifunctional reagents, 61
Heterogeneous immunoassay, xviii, 39, 55
Hexoestrol, 171–2, 177–8; *see also* Anabolic hormones
Homogeneous immunoassay, xviii, 55, 57, 75, 81, 147; *see also* Enzyme mediated immunoassay technique
Hook effect, xviii
Horse meat, detection of, 90–3, 95, 100–2, 104–5
Horseradish peroxidase, 56, 59–62, 90, 103, 142–3, 148
Horse serum albumin, 89
HPLC, 176–9, 189, 206
Hydroxyl groups, 201–2, 205–6, 208

Hydroxyquassins, 204–7
N-Hydroxysuccinimide ester, 61
Hygromycins, assay for, 20
Hypoallergenic infant milk formula, 115, 121

Immunisation schedules, 28–30, 41
Immunoadsorbent chromatography, 38, 47, 89, 93, 104; *see also* Affinity chromatography
Immunoanalysis; *see* Analysis, immuno-
Immunoassay, xviii
- amyloglucosidase, 126–38
- anabolic hormones, 171–80, 182–5
- diethylstilboestrol, 171–80
- meat species identification, 88–93, 96–102, 102–7
- milk protein denaturation, 116–21
- ochratoxin A, 162–5, 198–200
- potato glycoalkaloids, 190–4, 200–3
- quassins, 205–7
- quinine alkaloids, 207–10
- small molecules, 197–210
- soya protein, 111–13
- staphylococcal enterotoxins, 142–52
- 17α-trenbolone, 182–5

Immunochemiluminometric assays, 81
Immunodiffusion, 21, 87, 95–6, 116, 127, 131, 141, 148–50
Immunoelectrophoresis, 21, 87, 127, 129–30, 132
Immunofluorometric assays, 75
Immunogen, xviii, 26, 65, 88, 103, 161, 171, 190, 198–210
Immunoglobulin, xv, 42, 48, 96–7, 95–108, 116, 122, 125, 127, 142, 144, 147
- ammonium sulphate isolation, 96
- colostral whey, from, 97
- G, 99, 116, 122, 142, 144
- G, bovine, 99, 116, 122
  - as carrier protein, 161
- G, rabbit, 142
- rivanol isolation, 96

Immunometric assay, xviii, 21, 46–9, 55, 147–8
Immunoprecipitation techniques, *see also* Immunodiffusion; Immunoelectrophoresis, 21, 26
Immunoradiometric assay, 96
Immunoreactivity, xviii, 61–2, 74
Immunosorbent, for phase separation, 45
Imprecision, xx, 42, 45, 192
Inaccuracy, xviii
Incubation temperature, 65, 135
Incubation time, 31, 65, 133–8
Indirect radioimmunoassay; *see* Immunoradiometric assay
Infant formulas, cow's milk based, 115
Infectious diseases, 64–5, 67
Insects, 20
Iodogen, 25
Isobutylchlorocarbonate, 199, 206, 208
Iso-electric focusing, 88, 95, 106, 129
Isoluminol and its derivatives, 78, 80
Isoquassinic acid, 206–7

Kangaroo meat, influence of, 95, 100–1, 104–5
Keyhole limpet haemocyanin, 163
Kidney, porcine
- ochratoxin analysis, 165, 200
- stilbene analysis, 173

Kits, field testing, 54, 103, 107–8
Kits, immunoassay, 105, 107–8, 112

Label, xix, 25; *see also* Avidin; Biotin; Coenzymes; Enzymes as labels; Fluorophores; Free radicals; Luminescent labels; Particles; Phosphorescent probes; Radioisotopes; Viruses
α-Lactalbumin, 116–22; *see also* Milk proteins
β-Lactoglobulin, 116–22; *see also* Milk proteins
Lactoperoxidase, 25, 98
Lamb, detection of, 90–3; *see also* Sheep meat, detection of
Latex, 45
Law of Mass Action, 26–7, 64

Lectins, 38
Legumes, 20; *see also* Proteins, soya
Limit of detection; *see* Detection limit
Limit dextrins, enzyme hydrolysis, 125–6
Limited reagent immunoassay, xix, 22, 44, 182
Liver, stilbene analysis, 173
Luciferase, 78–9
Luminescence, 77
Luminescent labels, 25, 77–82
Luminoimmunoassay, xix, 25, 77–82
Luminol, 78, 80
Lysine, ε-amino groups, 61, 147, 199, 201
Lysozyme, 162

Magnetic particles, 40, 45, 63, 147–8
Matrix, xix, 38, 165, 169, 172, 176, 178
Meat
  adulteration, 87, 95
  anabolic hormones analysis, 169–85
  analysis of, 20
  analysis of mixtures, 87–93, 95–108
  bulked frozen, 95
  identification, 87–93, 95–108
  preparation of extract for analysis, 89, 96, 105
  speciation; *see* Meat, species identification
  species identification, 87–93, 95–108
    immunogens for, 103–4
Meat products, 111–13
  comminuted, 87, 105
  frozen, 95–109
  heat processed and cooked, 93, 112
  raw, unheated, 87–93, 95–109, 112
Merthiolate, 38, 62
Micro-ELISA plate; *see* Microtitre plate
Microperoxidase, 78
Microtitre plate, xix, 45, 54, 56, 59, 62–3, 66–7, 88–90, 98, 100–2, 116, 142, 147, 162, 182, 190–1
Milk
  cow's, 115–16
Milk—*contd.*
  heated, 116, 118, 121
  milk products, 20
  residual antigenicity after heat treatment, 115–16, 121–3
  skimmed, 115–16, 119
  unheated, 121
Milk protein
  denaturation, 115–23
  purified, 116–17
Milk proteins, 115–23
  cow's, 115, 121
    allergy to, 115
Misclassification, xix, 39
Mixed anhydride reaction, 61, 78, 163, 190, 199–200, 205–6, 208–9
Monoclonal antibodies, *see* Antibodies, monoclonal
Monovalent antigen, *see* Antigen, monovalent
Morphine, 23, 35
Multivalent antigen, *see* Antigen, multivalent
Muscle, stilbene analysis, 173, 175
Mushrooms, analysis of, 20
Mutton, detection of, 87; *see also* Lamb, detection of; Sheep meat, detection of
Mycotoxins, 159–65, 172, 198–200
  bioassays, 160–1
  chemical analysis, 161
  immunoassay, 161–5

Neoquassin, 204–7
Nigakilactones, *see* Quassins
p-Nitrophenylphosphate, 116, 127–8, 136
Noise factors, 37, 174–6
Non-competitive ELISA, xix, 49, 58, 128, 144–50, 182–5
Non-equilibrium immunoassay, xvii
Non-isotopic label, 46–67, 73–82, 162
  see also Avidin; Biotin; Coenzymes; Enzymes as labels; Fluorophores; Luminescent labels; Particles; Phosphorescent probes; Viruses

Non-precipitating antibody, 21
Non-separation assays, *see* Homogeneous immunoassay
Non-specific binding (NSB), xix, 37–8, 57, 66, 163, 172, 176, 178, 180–2, 184–5

Ochratoxins, 20, 159–65, 198–200
  bioassay, 160–1
  biological materials, in, 162–5
  chemical analysis, 161
  immunoassay, 161–5
  structures, 160, 199
Oestradiol, 170; *see also* Anabolic hormones
Oestrogens, 169–70; *see also* Anabolic hormones
On-site testing, 107–8

Parallelism, xix, 37, 165, 172, 176, 180
Particles, as labels, 74
Penicillium, 159, 198
Periodate oxidation, 61, 190, 201–2
Pesticides, assays for, 20
pH, effect of, 131, 133, 135
Phase separation, xix, 22, 27, 31, 39–46, 54, 62–3, 75, 147
  adsorption, 39–42
  chemical precipitation, 39–40
  chromatography, 39–40
  immunology, 39–40, 42–4
  solid phase, 39–40, 44–6
o-Phenylenediamine, 90
Phosphorescent probes, as labels, 77
Phylogenetic distinction, 49, 57, 93, 104
Pig meat, detection of, 87, 90–3, 100–1, 104–7
Plant cell cultures, 197, 203, 208
Plant metabolites, 203–10
Polyclonal antibodies, xx; *see also* Antiserum, polyclonal
Polyethylene glycol, 44, 63, 127, 134–7
Poly-L-lysine, 162
Polystyrene cuvettes for ELISA, 127, 131–2, 136
Pork, detection of, 90–3; *see also* Pigmeat, detection of
Potato, glycoalkaloids analysis of, 187–94, 200–3
Potency estimates, 36
Poultry, influence of meat from, 87
Precision, xx, 39, 42, 45, 64, 113, 118, 192
Precision profile, xx, 64
Principle of immunoassay, 22, 54
Protein A, 58, 96, 98, 101, 127, 144, 146
  enzyme-labelled, 58, 146
  ^{125}I-labelled, 96, 98, 101
  immunoadsorbent, 127, 148–9
  in sample, interference from, 146
  S. aureus, from, 144
Protein diluent for assays, 38, 98
Protein–enzyme conjugates, 61
Proteins, as analytes, 67, 75, 81
  amyloglucosidase, 121–39
  β-conglycinin
  detection of heat denatured, 115–23
  milk, 115–23
    denaturation, 115–23
  soya, 111–13
  soya bean, 120
  staphylococcal enterotoxins, 141–52

Quality control of assays, 24, 113, 118
Quassinoids, 204–7
  structures, 204
Quassins, 204–7
  structures, 204
Quenching, 24, 75–6, 172
Quinidine, 208–9; *see also* Quinine alkaloids
Quinine alkaloids, 207–10
  structures, 208
Quininone, 208–9: *see also* Quinine alkaloids

Rabbit meat, cross-reaction of, 92–3
Rabbits, for antiserum production, 31, 88, 90, 92, 97–9, 101, 103, 107, 116–17, 127, 129, 145–9, 161–3, 190

Radioimmunoassay, xx, 21–46, 53, 65–6, 73, 77, 81–2, 96, 98–102, 107–8, 141, 151–2, 171, 190
  anabolic hormones, for, 169–80
  capture, 102
  diethylstilboestrol, for, 171–80
  meat, species identification of, 96–102
  ochratoxin A, for, 161–2
  quinine, for, 207–10
  solanidine, for, 190, 193
  stilbenes, for, 171–80
  trenbolone, for, 184–5
  zeranol, for, 180
Radioiodination, 25–6, 46, 73, 98
  proteins, 98
  tags, 26, 161
Radioiodine, 24–5
Radioisotopes, 24–5, 53, 142, 161
Radiolabelled antigens, 24–6, 73, 161, 172–5, 190, 208–9
Rare earth metal chelates, 74, 77
Recovery determination, 175, 179–80
Residual antigenicity, after heat treatment, 115–16, 121–3
Rhodamine, 74–5
Rivanol, 96, 99
Rubijervine, 203; *see also* Glycoalkaloids
Ruggedness, xx

Salmonella, 20
Sample clean-up, bile, 173–81
Sample preparation,
  barley, 164
  beer, 128, 131
  bile, 176–80
  meat and meat products, 88–90, 96, 105, 173–5
  potato, 190
  soya products, 111
Sandwich assay, xx, 48–9, 54–8, 75, 77, 128, 144–52
  direct, 144–5
  indirect, 145–51
Sausage products, 113
Scatchard plot, 34
Scintillation cocktails, 24, 73
Sensitivity, xx, 27, 46, 53, 55, 64–6, 75–7, 81, 100, 102, 118, 131, 134, 138, 141–2, 148, 150, 152, 162, 171, 173, 192–3, 197, 199, 205, 207
  factors affecting, 64–5, 131–5, 137, 152
Serum albumin; *see* Albumin, serum
Sheep, for antiserum production, 31, 88, 146
Sheep meat, detection of, 90–3, 95, 100–1
β-Sitosterol, 203
Snake venoms, identification of, 108
Sodium azide, 38, 62
Sodium sulphate precipitation, 40
Soft drinks, analysis of, 20
Solanidine, 187–90, 193, 201–3; *see also* Glycoalkaloids
α-Solanine, 187–8, 190, 201–3; *see also* Glycoalkaloids
Solasodine, 203; *see also* Glycoalkaloids
Solid phase, 44–9, 55, 62–3, 66, 75, 77, 81, 102, 142, 145, 182
  assays, 81; *see also* Immunosorbent; Microtitre plate
Soya
  concentrate, 111, 113
    textured, 113
  extrudate, 111
  flour, 111, 113
  isolate, 111
  products, 111–13
  protein, detection of, 88, 111–13
Soyabean protein, 120
Species testing, 20; *see also* Analysis, immuno-; Meat species identification
Specificity, xxi, 33, 34–7, 107, 163, 197, 199
Standard curve, 23, 27, 54, 56, 59, 64–6, 91, 101, 118–19, 143, 145, 162, 171–2, 176, 178–9, 184–5, 191, 202, 206–8
Staphylococcal enterotoxins, 141–51
Staphylococcus aureus, 146
  toxins, 20, 141–3, 145–51

Starch, enzyme hydrolysis of, 125
Steric exclusion chromatography; *see* Chromatography
Steroids, 20, 62, 66, 81, 170–1, 176
Stigmasterol, 203
Stilbenes, 170–81; *see also* Diethylstilboestrol
Styrene-divinyl benzene, 177–8
Succinic anhydride, 206, 208–9
Sweetness, analysis for, 20
Synthetic hormones, 170

T-cells, 30
Testosterone, 170; *see also* Anabolic hormones
Thaumatin, assay for, 20
Therapeutic drug monitoring, 76–7, 169
Time resolved fluorimetry; *see* Fluoroimmunoassay, time resolved
Tissue purification procedures; *see* Sample preparation
Titre, xxi, 31–3
Tomatine, 193
Toxins,
  food, 198–203
  fungal, 20, 159–65, 198–200
  glycoalkaloids, 187–94, 200–3
  staphylococcal, 141–52
Tracer, xxi
Traditional analyte, definition, 6
17α-Trenbolone, 170–3, 177, 180–1, 184–5; *see also* Anabolic hormones
Trichinella, 20
Tricothecins, assays for, 20
Tritium, as a label, 24–5, 65–6, 73, 162, 171–2, 174, 178, 209
Tween 20, 90, 98, 105, 128, 134–5, 137
  effect of, 134–5, 137
Two-site assay, xx, xxi, 48–9, 102–7, 144

Umbelliferone, as a label, 74, 76
Undeclared meats, detection of, 87, 95
Urease, as a label, 103, 105, 152

Veal, detection of, 91, 93; *see also* Beef, detection of
Vegetables, analysis of, 20
Venison, influence of, 90, 92
Viral antigens, 67
Viruses,
  analytes, as, 20
  labels, as, 74
Vitamins, assays for, 20

Water of hydration, loss of, 44
Whey, 115–16, 118–20
  heat denatured, 115–16, 119–23
  pasteurised, 122
  protein, 112, 116, 121, 123

Zeranol, 170, 172–3, 177, 180–1; *see also* Anabolic hormones
Zeranolenone, 172; *see also* Anabolic hormones